PROTOPLASMATOLOGIA
HANDBUCH
DER PROTOPLASMAFORSCHUNG

HERAUSGEGEBEN VON

L. V. HEILBRUNN UND F. WEBER
PHILADELPHIA   GRAZ

MITHERAUSGEBER

W. H. ARISZ - GRONINGEN · H. BAUER - WILHELMSHAVEN · J. BRACHET - BRUXELLES · H. G. CALLAN - ST. ANDREWS · R. COLLANDER - HELSINKI · K. DAN - TOKYO · E. FAURÉ - FREMIET - PARIS · A. FREY - WYSSLING - ZÜRICH · L. GEITLER - WIEN · K. HÖFLER - WIEN · M. H. JACOBS - PHILADELPHIA · D. MAZIA - BERKELEY · A. MONROY - PALERMO · J. RUNNSTRÖM - STOCKHOLM · W. J. SCHMIDT - GIESSEN · S. STRUGGER - MÜNSTER

BAND II
CYTOPLASMA

C
PHYSIK, PHYSIKALISCHE CHEMIE, KOLLOIDCHEMIE

1
THE VISCOSITY OF PROTOPLASM

WIEN
SPRINGER-VERLAG
1958

THE
VISCOSITY OF PROTOPLASM

BY

L. V. HEILBRUNN
PHILADELPHIA

WITH 23 FIGURES

WIEN
SPRINGER-VERLAG
1958

ISBN-13: 978-3-211-80485-8   e-ISBN-13: 978-3-7091-5458-8

DOI: 10.1007/978-3-7091-5458-8

The Viscosity of Protoplasm

By

L. V. HEILBRUNN

Zoological Laboratory, University of Pennsylvania,
Philadelphia, Pennsylvania

With 23 Figures

Table of Contents

Introduction

From the very beginning of the study of protoplasm, biologists and cell physiologists have been concerned with the attempt to understand as much as possible about its physical state. But it is doubtful if in any other field of biology there has been more erroneous information gathered and perpetuated. This has in part been due to a lack of understanding of basic physical facts, in part also to the publication of numerous careless and untrustworthy observations. Microdissectionists prodded the cell with their microneedles, and they made many definite statements both concerning the relative and also the absolute viscosity of the protoplasm they attacked. These observations were completely subjective and were in addition handicapped by the fact that when a needle or other foreign object enters a cell, it is almost certain to cause a clotting of the protoplasm in its immediate vicinity. One observer saw no Brownian movement in the protoplasm of marine egg cells and concluded from this that the viscosity of the protoplasm of these eggs must be as great or greater than that of glycerol, in which he assumed no Brownian movement possible. And this in spite of the fact that numerous observers have noted vigorous Brownian movement in such eggs, the amplitude of the movement being limited by the fact that the cell is so clogged with granules that the distance any one granule can move is restricted. Also one can scarcely draw conclusions as to viscosity merely from the presence or absence of Brownian movement; indeed, Brownian movement can occur in materials even as solid as glass. Muscle physiologists have at times made measurements of the "viscosity" of an entire muscle fiber, outer membrane, cortex and interior all considered as one material, or they have even measured the viscosity of an entire muscle. One might as well make viscosity measurements of a solution and include in the value obtained the viscosity of the bottle which contained it.

In the present discussion, no attempt will be made to evaluate or to examine critically all the older observations which served only to becloud the literature. In a monograph published nearly thirty years ago, the limitations, the inaccuracies, the absurdities of some of these old observations were clearly pointed out (Heilbrunn 1928), and for the most part, they are no longer given the consideration they once were. Although, unfortunately, once a false statement finds its way into the literature, it is never completely forgotten and careless commentators often continue to refer to it.

At the present time, there is a rather large and a consistent body of information concerning the viscosity of protoplasm, and this information is based on correct methods of study. Fortunately, it is possible to make viscosity measurements of the protoplasm of living cells without injuring them. The knowledge that has been gained can be used to throw light on some of the most puzzling problems the cell physiologist has to face. The living cell is a colloidal engine; often as in the case of the muscle fiber, it can do mechanical work. In order to understand such an engine, it is essential to know as much as possible about its physical make-up. If we are to understand the colloids which constitute the engine, we must have some way of knowing what they are like when the engine is in a state of comparative rest and how they change when it becomes active. In the search for such information, the study of viscosity is of prime importance.

Moreover, it is essential that our knowledge be gained directly by a study of living protoplasm. In the past—and indeed in the present as well—chemically-minded biologists and physiologically-minded chemists have sought to interpret the colloidal behavior of protoplasm on the basis of what one might deduce from the presumed colloidal behavior of the substances of which it is composed. But, as was clearly shown in the monograph already referred to, the colloidal behavior of protoplasm can not be deduced from any knowledge of the colloidal behavior of proteins alone or of proteins mixed with lipids. The very fact that protoplasm is alive should *a priori* indicate that its physical behavior must be different from that of lifeless materials, and so indeed it is. Very different, indeed. Nor is it helpful, as some biochemists have thought, to purify proteins from living tissues and then from studies of the colloidal properties of these purified proteins to draw conclusions as to the colloidal properties of the tissues from which they were derived. For our knowledge of colloid and polymer chemistry has demonstrated over and over again that the physical behavior of colloids and/or polymers is markedly conditioned by the impurities they contain, that is to say, by traces of substances which can have a profound effect. Also if we are to attempt to study model systems rather than the protoplasm itself, we must know how to construct our models and this information can best be obtained from a knowledge of the colloidal behavior of the living material.

The Viscosity Concept

As everyone knows, liquids vary greatly in the ease with which they flow. The more readily a liquid flows, or the more readily its particles move over each other, the greater its fluidity. And because viscosity is the inverse of fluidity, the greater the fluidity of a liquid, the less its viscosity. Viscosity can be defined as internal fluid resistance. It is the resistance offered by one portion of a liquid flowing over another portion of the same liquid. In order to compare the viscosity of various liquids, we need some sort of a coefficient or unit. In order to arrive at such a unit, we may consider two parallel liquid surfaces of unit area and unit distance apart. If the liquid is to have unit viscosity, unit force must be required to cause the two liquid surfaces to move past each other at a velocity of one cm. per sec. Viscosity is then force per unit area divided by velocity per unit length. In c. g. s. units this becomes $M\,l^{-1}\,t^{-1}$. In other words, for a liquid of unit viscosity, one dyne is necessary to cause planes of the liquid of unit area and unit distance apart to move over each other at unit speed.

The unit of viscosity, defined as above, is the poise. Water, at a temperature of 20^0 C., has a viscosity almost exactly one-hundredth of a poise. Hence it is convenient to use the term centipoise and to consider this as equal to the viscosity of water (at 20⁰ C.). Some liquids, for example, methyl alcohol, ether, acetone, chloroform and benzene are more fluid than water and have a viscosity less than one centipoise. On the other hand, ethyl alcohol is more viscous than water, sulfuric acid has a viscosity of approximately 27 centipoises, many oils have a viscosity of about 100 centipoises, and glycerol at 20⁰ C. has a viscosity of 830 centipoises. In all these cases, the viscosity does not vary with the applied force, and such liquids, inasmuch as they obey Newton's law for fluids, are called "Newtonian."

Ordinarily, when substances are dissolved in water or other solvents, the viscosity of the resultant solution is not very different from that of the pure solvent. But this is certainly not true for substances of high molecular weight, that is to say for macromolecules or polymers. The presence of these substances in solution may have a marked effect on the viscosity of the solution, and the effect is especially pronounced if the macromolecules are non-spherical. In such solutions of macromolecules, or in colloidal solutions, it not infrequently happens that the viscosity varies with the force applied to produce flow (often called the shearing force). Thus colloidal solutions may be non-Newtonian in their behavior. Consider gelatin, for example. Fairly high concentrations of gelatin may be obtained by dissolving this protein in solutions of warm water; then as the solution cools it tends to become stiff, that is to say, gel.

A solution of gelatin which is in the gel state or is about to transform into a gel has a high viscosity under ordinary conditions. Indeed, if the gel is reasonably firm, it will not flow at all under the influence of gravity. But under higher pressures, the gelatin gel can be made to flow, and indeed its viscosity may become rather low. When, as in this instance, viscosity varies with shearing force, it is customary now to speak of an "apparent viscosity" or an "anomalous viscosity" (see Barr 1931, Scott Blair 1938, 1949). Often the change in viscosity is not so great or so apparent as it is in the case of a gelatin gel, and it is necessary to make proper measurements of the viscosity in order to determine how it varies with the shearing force. The phenomenon of decrease in viscosity with higher shearing forces is called thixotropy, and materials which exhibit this phenomenon are called thixotropic. But viscosity does not always decrease as the shearing force is increased. Sometimes exactly the opposite occurs. This was first observed in dense suspensions of quartz particles, as in sand. For such suspensions, as the applied force increases, the viscosity of the suspension as a whole increases. This is called dilatancy, and the materials showing it are called dilatant. As will be noted later, both thixotropy and dilatancy occur in various types of protoplasmic systems.

All protoplasm contains granules visible with the light microscope, and these granules may be extremely numerous; so numerous in fact that there is scarcely any space left between the individual granules. This is the condition found in paramecium, as well as in many other cells. Actually, as can be shown mathematically, if the granules are spherical and of the same size they can not occupy more than 74.05% of the total volume of the cell. Such a high concentration, or even one approaching it, represents a highly concentrated suspension. It is necessary, therefore, to present a brief discussion of the facts that are known concerning the viscosity of suspensions, and especially the viscosity of highly concentrated suspensions.

EINSTEIN (1906 a, 1911, 1920) was the first to develop a theory for the viscosity of suspensions. He arrived at a formula for the viscosity of a suspension in which the suspended particles are solid spheres, large in size in comparison with the molecules of the dispersion fluid, but still relatively small in comparison with the apparatus containing them. This formula is:

$$\eta_s = \eta_0 \,(1 + 2.5\,c),$$

in which η_s is the viscosity of the suspension, η_0 the viscosity of the dispersion medium, and c the concentration of suspended particles (that is to say, the ratio of the volume of the suspended particles to the total volume of the suspension).

Although this equation is reasonably adequate for dilute suspensions, provided that the suspended particles are solid and spherical, it does not hold for concentrated suspensions, for in his mathematical analysis, EINSTEIN did not take into account an interaction between the suspended particles; EINSTEIN himself emphasized the limitations of his equation. Following EINSTEIN, various authors have suggested equations which might hold for concentrated suspensions. So, for example, there is the equation of SIMHA (1949, 1950), according to which,

$$\eta_s = \eta_0 \,(1 + a_1\,c + a_2\,c^2 + a_3\,c^3 + \dots).$$

When c is small, this equation reduces to the EINSTEIN equation, for then c^2 and c^3 become negligible. The value of a_1 can be taken as 2.5, but for a_2, different authors prefer different values, and these may vary from 7 to 14 (see OVERBEEK 1952). One thing is certain—in concentrated suspensions the viscosity is much higher than that which would be calculated on the basis of the simple EINSTEIN equation. Another important fact is that concentrated suspensions tend to show the phenomenon of dilatancy. This is certainly true of concentrated suspensions of quartz, or of sand, or of starch paste.

The Measurement of Protoplasmic Viscosity

As the great physiologist LUDWIG once so aptly said, "Die Methode ist alles." Nowhere is this epigram more important than in the study of the viscosity of protoplasm. If we are to have adequate information concerning the viscosity of protoplasm, we must have proper methods for measuring it.

A proper method is one which gives objective information, preferably quantitative. It must be a method that does not destroy the life of the protoplasm it measures; and it should cause as little injury as possible.

preferably no injury at all. The methodology of protoplasmic viscosity measurements has been discussed at some length by Heilbrunn (1950).

Ordinarily when a physicist or a physical chemist measures viscosity, he uses an Ostwald viscometer, or some modification of it. With this simple apparatus it is possible to determine the length of time it takes for a given volume of liquid to flow through a capillary tube. The pressure pushing the liquid through the tube is simply gravity. According to Poiseuille's law, the volume of liquid that can pass in a given time varies as the fourth power of the radius of the tube, and inversely as the length of the tube and the viscosity of the liquid. For non-Newtonian liquids it is necessary to have a viscometer in which the liquid is forced through the capillary at varying pressures. Unfortunately, this method of measurement can not be used for living, non-injured protoplasm, for if the protoplasm of the interior of a cell is to be pushed through a tube, it must be taken out of the cell, and this under ordinary conditions destroys the life of the protoplasm. Actually, what happens is that when protoplasm emerges from a cell it clots in much the same way as blood clots, in a reaction or series of reactions the first stages of which require the presence of calcium ion (see later discussion). But just as it is possible to measure the viscosity of blood by preventing the clotting by the addition of oxalate or citrate, it might conceivably be possible to get relatively normal protoplasm out of cells by crushing them in the presence of oxalate or citrate and then observing the rate of flow of this decalcified protoplasm through a small capillary. Pfeiffer (1939, 1940) did force small droplets of protoplasm through a glass capillary, and although he made no attempt to prevent clotting, he was able to obtain low values (4--5 centipoises) for the viscosity of the protoplasm of a slime mold (*Phycomyces* sp.) and two algae (*Chara* and *Vaucheria*). This is rather surprising, for according to Kamiya (1942), "It is impossible to make slime mold protoplasm flow through a capillary without fatal results."

In cells in which protoplasm normally flows, as in ameba, one could apply Poiseuilles's law if one knew the pressure back of the movement, for one can easily obtain information as to the dimensions of the tube through which the protoplasm is flowing. Perhaps in a large ameba it might be possible to apply pressure to one end and then observe the rate of flow. Something of the sort was done by Kamiya (1940, 1942). He applied gaseous pressure to bits of protoplasm taken from slime molds. However, in these experiments there is uncertainty as to the effect the rigidity of the outer membrane of the droplet might have. Pressure applied to this outer membrane might not be passed on in its entirety to the interior.

Clearly, the common method of measuring viscosity by studying the flow of a liquid through a tube is not applicable to normal living protoplasm inside of cells. Fortunately, there is another widely used and thoroughly acceptable method which can be applied to living protoplasm. This method depends on the fact that the rate of movement of an object through a fluid depends on the viscosity of the fluid. In the Hoeppler viscometer, which can be obtained in various forms from apparatus companies, steel or glass balls are allowed to fall through tubes containing liquids. With this viscometer. fluids with viscosities varying from 0.6 to 100,000 centipoises can be accurately measured.

The method depends on the application of STOKES' law, a law derived by the mathematician STOKES over a hundred years ago, and now widely used both by physicists and physical chemists. In its original form, the law states that when a sphere is pushed through a liquid, its speed v varies directly with the propelling force W, and inversely with the radius of the sphere a and the viscosity of the fluid η. Thus,

$$W = 6\,\pi\,\eta\,a\,v.$$

If the sphere falls under the influence of gravity, then the force acting on it is its apparent weight times the gravity constant g. The apparent weight is the volume of the sphere times the difference in its specific gravity σ and the specific gravity of the fluid ϱ. (If the sphere is lighter than the fluid surrounding it, the sphere will of course rise rather than fall.). Substituting proper values for the force of gravity, that law becomes

$$^4/_3\,\pi\,a^3\,(\sigma - \varrho)\,g = 6\,\pi\,\eta\,a\,v,$$

and solving for η,

$$\eta = \frac{2\,g\,(\sigma - \varrho)\,a^2}{9\,v}$$

Thus if we have a sphere of known radius and known specific gravity falling through a liquid of known specific gravity, we can readily determine the viscosity by measuring the speed of fall. And if c. g. s. units are used throughout, the equation gives the viscosity in poises.

But there are certain limitations to the use of STOKES' law. In his original derivation of the law, STOKES assumed that certain mathematical terms, the so-called semi-quadratic terms, could be neglected. But this is only true if the movement of the sphere is slow and its size small. According to RAYLEIGH (1893), this assumption is warranted if $\frac{v\,a\,\varrho}{\eta}$ is negligible as compared to unity, and modern authors agree that the original form of STOKES' law holds if the REYNOLDS number $\left(\frac{2\,v\,a\,\varrho}{\eta}\right)$ is small. BARR (1931) states that if the REYNOLDS number is less than 0.05, STOKES' law is accurate to about 1% (see also SCHILLER 1932). When the REYNOLDS number is larger, a more exact formulation has been derived by OSEEN. However in studies that have been made on protoplasm, OSEEN's law does not need to be used, for in general in studies of protoplasmic viscosity, the REYNOLDS number is extremely small. Thus for the protoplasm of the egg of the sea urchin *Arbacia*, the REYNOLDS number is of the order of the magnitude of 3×10^{-6}.

There are also a number of other assumptions involved in the derivation of STOKES' law. The motion of the sphere must not only be slow, it must be steady and free from acceleration, there must be no slip between the sphere and the fluid, the sphere must be rigid, and the fluid through which it moves must be homogeneous and extend infinitely in all directions.

Arnold (1911) was able to show that these assumptions do not affect the validity of the law for a small sphere dropping through a liquid. If the sphere falls through a narrow tube, a correction, the Ladenburg correction, (Ladenburg 1907) is necessary, but according to Bingham (1922) this correction becomes unimportant if the diameter of the tube is more than ten times that of the sphere.

Ordinarily, the particles which are present as inclusions in protoplasm do not fall readily under the influence of gravity, or rather because of their small size, they fall so slowly that in most instances their fall tends to be counteracted by the Brownian movement of the particles, a movement which tends to distribute the particles through the surrounding medium. Hence in order to make viscosity measurements, it is necessary to subject the particles to a force greater than gravity. Then in using Stokes' law it is necessary to substitute for the gravity constant g, a term $c\,g$, in which c is the centrifugal force in terms of gravity.

Stokes' law was derived for a single sphere. But in cytoplasm the small spherical granules are present in large numbers. If then one is to determine the viscosity of the cytoplasm, it is essential to have some sort of a correction for Stokes' law. A suitable correction was proposed by Cunningham (1910), and when this is introduced into the equation for the viscosity given previously, this becomes

$$\eta = \frac{2\,g\,(\sigma - \rho)\,a}{9\,q\,v}$$

in which

$$q = \frac{4\,b\,(b^5 - a^5)}{(b-a)^2\,(4\,b^4 - b^3\,a - 6\,b^2\,a^2 - b\,a_3 - 4\,a^4)}$$

In this equation, a is the radius of one of the sphere; b is half the distance between adjacent spheres. Morever, in making the computation, Cunningham found it better to substitute for b a quantity b' which is equal to $1.5\,b$. When the number of granules in the cytoplasm is relatively small and the granules are relatively far apart, Cunningham's correction becomes of less importance. Moreover, in most instances the student of protoplasmic viscosity is primarily interested in changes in viscosity which occur during vital processes or as the result of one sort of treatment or another, and in such measurements, the Cunningham correction need not be considered unless the size or number of the granules changes markedly or the distance between the granules changes as a result of an increase or decrease in the volume of the entire cell.

Practical details of technique involved in the use of the centrifuge methods are discussed by Heilbrunn (1950). In general a hand centrifuge is preferable to an electric centrifuge. Certainly this is true in those instances in which granules move through a cell after only a few seconds of exposure to centrifugal forces of low intensity. An appreciable time interval is required before an electric centrifuge reaches a given speed and also there is a time interval before the centrifuge can be stopped. The speed of the moving granules is determined by centrifuging cells for varying lengths of time; in this way it is possible to discover how long it takes for the granules to travel a given distance. With a microscope centrifuge such as that described by HARVEY and LOOMIS (1930; see also HARVEY 1932 b), it might be possible to measure the movement of individual granules while they were being centrifuged, but this has never been done, or at any rate no such data have ever been published. This is in part due to the fact that in many types of protoplasm the granules move so rapidly under weak centrifugal forces that they have already moved as far as they can go before they can properly be observed with the microscope centrifuge. Another difficulty lies in the fact that with HARVEY's apparatus it is often difficult to keep cells in a field long enough to observe them properly. This difficulty is overcome in the Danish microscope centrifuge manufactured by Struers in Copenhagen.

In order to use STOKES' law, it is necessary to know the specific gravity of the granules as well as the specific gravity of the fluid through which they move. The specific gravity of the granules can be measured by removing them from the cell and centrifuging them in various concentrations of sugar solution. It is not too difficult to determine that concentration of sugar in which the granules move neither to the top or the bottom of the centrifuge tube.

The question has sometimes been raised as to whether centrifugation may not injure the protoplasm in such a way as to affect the viscosity. This is perhaps possible in some instances. But many living cells can withstand enormous centrifugal forces, forces far above those ordinarily used in the determination of protoplasmic viscosity. So for example heart cells from embryo chicks when raised in tissue culture can continue to grow and to pulsate after exposure to a centrifugal force of 400,000 times gravity for half an hour (MACDOUGALD, BEAMS and KING 1937), and *Ascaris* eggs can withstand 400,000 times gravity for an hour. Eggs of the sea urchin *Arbacia* centrifuged at 8 or 10 thousand times gravity so that the granules in the interior protoplasm are completely shifted will develop perfectly. In the *Chaetopterus* egg, it is only when the granules in the stiff outer cortex of the egg are moved that development is influenced (WILSON and SCHUB 1953); on the other hand, centrifuging the granules in the interior of the cell is without effect on the development (LILLIE 1908). If centrifugal force were to have an effect in changing protoplasmic viscosity, then it would almost certainly be true that greater centrifugal forces would cause more marked changes than lower forces; and yet in *Arbacia* eggs, according to HOWARD (1932), this is not true. HEILBRUNN (1926 a) also found a constancy of viscosity values for various centrifugal forces when these were used to determine the viscosity of the protoplasm of the egg of the clam *Cumingia*.

Another possible difficulty with the centrifuge method is the fact that as conditions in the cell change, or as the cell is exposed to some physical or chemical agent whose effect is being tested, the size and/or the specific gravity of the granules may change. This is undoubtedly a possibility and it must always be kept in mind. Occasionally it has not been considered when it should have been. But ordinarily the difficulty is not serious. For in most types of protoplasm, there are granules which are heavier than the hyaline protoplasm and those which are lighter. In general, any condition or influence which causes the heavier granules to move more slowly or more rapidly will have the same effect on the lighter granules and it is hard to conceive of any agent which would simultaneously cause one sort of change in the specific gravity of the heavy granules and exactly the opposite sort of change on the specific gravity of the light granules. And unless such opposite changes are postulated it is not possible to account for the fact that the viscosity as determined by movement of the heavy granules is the same as that determined by movement of the light granules.

Various botanists have used the centrifuge method to determine the viscosity of the protoplasm of plant cells. When marine egg cells or the cells of protozoa are centrifuged, the cells as a whole immediately pass to the bottom of the centrifuge tube. But when plant cells are centrifuged, conditions may be quite different. So, for example, if an *Elodea* leaf is centrifuged, the cells can not move from the rigid cell walls that contain them. Within a typical plant cell, there is a relatively thin layer of protoplasm surrounding a large central vacuole. When such cells are centrifuged, the entire mass of the protoplasm may break away from the cell wall and become piled up in a heap at the centrifugal end of the walled chamber in which it is enclosed (Jungers 1934, Beams 1949, Breckheimer-Beyrich 1949). Such an effect to a large extent defeats the purpose of the experiment. However, in many cases the protoplasmic layer retains its original position, and one can then determine the rate of movement of chloroplasts, starch grains or other inclusions through the protoplasm. Some plant cells have large chloroplasts which occupy a considerable volume of the cell and appear to be anchored to the outer cortex of the protoplasm. This is the condition in the case of the well known alga *Spirogyra*, the cells of which have large spiral chloroplasts. Under the influence of centrifugal force, these may be torn loose from their anchorage in the cortex and hurled to one end of the cell. Such an effect was observed by Weber (1921) and by various subsequent observers (see for example Northen and Northen 1938). Northen (1936) also made similar observations on cells of the related alga *Zygnema*. The ease with which chloroplasts like those of *Spirogyra* are cut loose from their moorings and move through the cell is not just a measure of the viscosity of the interior protoplasm of the cell, but to a large extent also a measure of the rigidity of the cortex or outer membrane to which they are attached. Small wonder then that from his studies Northen concluded that protoplasm was elastic. Indeed, the various studies of Northen (1938 a, b, 1939, 1940, Northen and MacVicar 1939), based as they are on the ease with which chloroplasts are torn loose and then move, are really more than anything else determinations of the rigidity of the cortex and should be considered as such. This was pointed out by Heilbrunn (1943, 1952). Also the rather variable degree of rigidity of the chloroplasts is a factor to be considered (see Eibl 1939).

In all types of physical measurements, it is advantageous to have corroborative evidence based on a variety of techniques, and this is especially true of measurements made on biological material, for in the study of living cells the chances of error are indeed great. Fortunately in addition to the gravity and centrifuge methods, there is a second method of viscosity determination which can be used on living cells. This method depends on the quantitative study of Brownian movement in the protoplasm.

The fact that Brownian movement can and does occur in living cells has been known for a very long time. Indeed, long before Brown made his famous discovery in 1828, the movement he saw was well known. But it was thought to be a characteristic of life, and it was Brown who showed that it could exist as well in inanimate materials. It is not certain who it was who first described Brownian movement of granules or particles within the protoplasm of living cells, but soon after the middle of the nineteenth century many observers described such movement and some of them thought that the speed of the movement was an index of the fluidity or viscosity of the protoplasm. However, these early observations gave no quantitative measure of viscosity nor could they have done this until the theory of Brownian movement had been worked out. Not until EINSTEIN (1905, 1906 b) and von SMOLUCHOWSKI (1906) had derived proper equations was it possible to use the speed of Brownian movement as any accurate measure of viscosity.

EINSTEIN's formula as given in various books of physical chemistry is as follows:

$$D_{x^2} = \frac{R}{N} \cdot \frac{T\,t}{3\,\pi\,\eta\,a}$$

in which D_x is the displacement of a particle along one axis in the time t and at the absolute temperature T, R is the gas constant, N the Avogadro number and η the viscosity. Using a value of 6×10^{23} for the Avogadro number and 8.3×10^7 for the value of the gas constant in ergs, this formula can be expressed as follows:

$$D_{x^2} = 14.7 \times 10^{-18} \cdot \frac{T\,t}{\eta\,a}$$

VON SMOLUCHOWSKI's formula is only slightly different—in it the value for D_{x^2} is $\frac{32}{27}$ times as great as it is in the EINSTEIN equation. Neither of the formulae is very exact, for various mathematical simplifications were involved in their derivation, and it was also assumed that the particles were rigid spheres and that the forces of interfacial tension were negligible. Hence although there is a fairly good agreement between calculated and observed values for D_{x^2}, the agreement is by no means close.

In living cells containing a large number of closely packed granules, these are very much restricted in their Brownian movement, and it is not possible to use the EINSTEIN formula. Nor can the formula be used if the granules are close to a restraining surface which tends to impair their free

motion. However, even when the granules are numerous, it is possible to
centrifuge them to one side of the cell and then observe their return. One
fortunate fact is that in using the Einstein equation it is not necessary to
know the specific gravity either of the granules or of the fluid through
which they move.

In some cells the granules are sufficiently far apart and are far enough
from bounding membranes so that it is possible to obtain a measure of the
viscosity by observing the Brownian movement. In such cells Pekarek
(1930 a, b, 1931, 1932) has made what appear to be proper measurements
of protoplasmic viscosity. Instead of using the Einstein equation, he uses
one derived by the physicist Fürth (1930):

$$\eta = \frac{R}{N} \cdot \frac{T\,t}{3\,\pi\,a\,l^2\,n}.$$

In this equation, most of the symbols are as before, l is a given distance
to the right or left of the original position of a particle in Brownian
movement, n is the number of times this distance is traversed in the
time t. Pekarek tested his method on distilled water and found excellent
agreement with the known value for the viscosity. However, since Pekarek's work some 25 years ago, scarcely anyone has attempted to use the
method. This is doubtless due to the fact, as Pekarek himself points out,
that the method can not properly be used if the granules in the protoplasm are rather close to each other.

There is also an interesting method of viscosity determination which can
be used when protoplasm is present in large masses. In slime molds just
before fruiting the protoplasm is not in the form of strands but is in a large
undifferentiated mass. Into such slime molds, Heilbronn (1922) inserted
small iron rods, and then he determined the number of amperes which had
to be passed through a nearby electromagnet in order to twist the rods.
The individual rods were 220 microns long and 50 microns in diameter.
The number of amperes necessary to twist a rod imbedded in protoplasm
was divided by the number of amperes necessary to twist it in a drop of
water and this was taken as the absolute viscosity in terms of water, that
is to say in centipoises. Heilbronn tested his method on drops of liquids
of known viscosity. He found it to be reasonably accurate, but always the
method gave higher values for small drops than for large ones, and thus
for smaller masses of protoplasm, the viscosity values were apt to be too
high. Also it seems clear that when a foreign object such as an iron rod
is inserted into protoplasm, it will immediately become surrounded by a
protoplasmic membrane or film, and such a film might have an influence
in retarding the free movement of the rod. This effect would perhaps not
be important for a relatively large object such as the rods used by Heilbronn, for the amount of metal acted on by the magnetic field would be
very large compared to the protoplasmic membrane, but it would probably
be important for small amounts of magnetic material taken up by a cell
in the form of vacuoles. Thus in the studies of Crick and Hughes (1950)
on fibroblasts, the iron particles taken up by these cells in the course of

their phagocytic activities would be surrounded by vacuolar membranes which would retard their free movement under the influence of magnetic forces. This fact was realized by CRICK and HUGHES.

Finally, there is a method which although it does not provide a quantitative measurement of viscosity, does give objective information concerning viscosity change. This is the plasmolysis method of WEBER (1924 a, b, c, 1925 a, b, c, d, 1929 a). It has frequently been used by botanists for it is relatively simple and it can be used on types of cells which do not lend themselves to study by other methods. When plant cells are plasmolyzed, the border of the cell as it leaves the cell wall may be perfectly smooth, in which case the plasmolysis form is said to be "convex," or it may be angular ("eckig"). The convex type of plasmolysis is an indication of relatively low protoplasmic viscosity, whereas when the viscosity is higher, the plasmolysis takes on an angular form. Also with lower viscosity, plasmolysis occurs more rapidly. It is thus possible to compare not only the form of the plasmolysis, but also the time required for the process to take place. Obviously the method is an indirect one, and there may be factors other than the viscosity which could influence the form the plasmolysis takes. Thus WEBER himself believes that the stickiness or adhesiveness of the outer layer of the protoplasm might exert an influence. He also states that the information the method gives concerns especially the outer layers ("Grenzschichten") of the protoplasm. A similar opinion has recently been expressed by GÜNTHER (1957). On the whole, the method has proven useful and has given results in the main consistent with the results obtained with more direct methods.

Absolute Viscosity

At the present time our information concerning protoplasmic viscosity is rather fragmentary. Only certain types of cells can be used for quantitative measurements. So, for example, if a cell has a very dense suspension of granules these can not be moved by centrifugal force. However, in such cases it should be possible to follow the movement of the nucleus through the cell. Most of the knowledge we now have concerns isolated cells—protozoa, egg cells, etc. Cells that are organized in tissues are more difficult to study. And yet, there are ways of separating tissues into their individual cells, and it is also possible to centrifuge whole organs or organisms and section them later, so that in the future we should have data concerning the viscosity of the protoplasm of cells from the liver, spleen and other organs.

In what follows, an attempt will be made first to survey our present knowledge concerning the absolute viscosity of the protoplasm of different cells. Following this, there will be a discussion of how the viscosity of the protoplasm changes during activity and as a result of exposure to various agents. One point that must be kept constantly in mind is the fact that the viscosity of the protoplasm of all parts of the cell may not be the same. Indeed, there is an abundance of evidence that in many cells

—perhaps in all cells—there is an outer cortical region or cortex which is much less fluid than the interior. In some cells this cortex is in the form of a thin membrane; in other cells it may be relatively thick. The physical properties of the cortex are being considered in another part of this treatise (II, E, 1) by Walter L. Wilson, so that in this section we shall primarily be concerned with the protoplasm in the interior of the cell. However, for some of the data in the literature, especially the data on plant material, it is hard to decide whether the measurements concern the cortex, the interior protoplasm, or both. Also our knowledge of what is happening to the interior is sometimes dependent on the changes that are occurring in the cortex. Hence it will be necessary at times to refer to the measurements that have been made of the viscosity of the protoplasm or the cortex. It should be noted that the thickness of the cortex may vary during the life of a given cell, and it is quite possible or even probable that at times the cortex may become so thick that it will come to occupy a very appreciable fraction of the entire mass of the protoplasm. In some cells certainly, cortical protoplasm can change into protoplasm of the interior, and vice versa. Conditions may vary in closely related types of protoplasm. Thus, in various species of amebae, the cortex may be very thin or it may constitute a sizable fraction of the entire cell, a fraction which may be highly variable. This changing relation between cortex and interior, a condition perhaps most evident in the giant ameba *Chaos chaos*, can introduce a serious complication in the interpretation of viscosity measurements. This will be discussed more fully in the section on Protozoan Protoplasm.

Plant Protoplasm

Writing in 1921, the botanist Meyer deduced that protoplasm must have a viscosity greater than that of castor oil, and he gives the viscosity of this oil as 1,060 centipoises. The idea that protoplasm in general has a viscosity of about a thousand times that of water has been fairly widespread and indeed writers of elementary textbooks of botany are prone to emphasize the high viscosity of protoplasm.

And yet, as early as 1903, Ewart, after stating it was "obviously impossible to measure the viscosity of protoplasm directly," concluded on theoretical grounds that there were "weighty reasons for considering the viscosity of the main bulk of the streaming protoplasm [of plant cells] to lie within the limits of $\eta = 0.04$, and $\eta = 0.2$ at 18^0 C." In other words, such protoplasm was assumed to have a viscosity between 4 and 20 centipoises. And in his calculation of the energy involved in the streaming of protoplasm in a *Nitella* cell, Ewart uses a value of 7.5 centipoises.

The first worker to make anything like a quantitative measurement of the viscosity of protoplasm was the German botanist Heilbronn. In 1914 he studied the rate of fall of starch grains through protoplasm. In order to obtain cells to study, he made sections through the upper part of the stem of the broad bean plant (*Vicia faba*). In these plant stems, there is an endodermis layer the cells of which contain starch grains—for this reason the layer is often called the starch sheath. Heilbronn measured the speed of

fall of the starch grains in the protoplasm and then he removed some of
the grains and measured their speed of fall in water. On the assumption
that the speed of fall was an inverse measure of the viscosity, HEILBRONN
divided the speed in protoplasm by the speed in water and regarded this
as a measure of the viscosity of the protoplasm in terms of the viscosity of
water. In this way, he arrived at values which varied from 8 centipoises to
infinity. Excluding the infinite values, the average was 24 centipoises.

These values obtained by HEILBRONN should not be regarded as correct
values for the viscosity but rather as maximum values. For one thing, in
order to obtain cells for observation, HEILBRONN had to cut sections. The
injury caused by such sectioning can markedly increase the protoplasmic
viscosity and can even induce gelation (WEBER 1924 a, BÜNNING 1926). The
fact that HEILBRONN occasionally obtained infinite values for the viscosity
can easily be explained on this basis. Secondly, HEILBRONN should have
used STOKES' law instead of merely comparing the speed of fall of the
starch grains in protoplasm and water, for the specific gravity of the protoplasm is not the same as that of water. (The first to realize that the use of
STOKES' law was essential was WEBER in 1917.) Then too, the layer of protoplasm through which HEILBRONN watched the starch fall was presumably
a very narrow layer, for the cells he observed consisted for the most part
of a large vacuole, and indeed HEILBRONN had to be careful to find grains
which were not too close to the outer wall of the cell. The presence of a
wall in the vicinity of the falling grains can cause a serious error (see
previous section), and in order to obtain proper values it would be necessary to introduce the LADENBURG correction to STOKES' law. Finally, the cells
that HEILBRONN studied are presumably crowded with starch grains and
they contain other inclusions as well, so that for a correct determination of
the protoplasmic viscosity in absolute units, it would be necessary to
introduce a correction such as the one proposed by CUNNINGHAM.

But the important point to be emphasized is that HEILBRONN found
values for the protoplasmic viscosity which were much lower than had
been assumed by earlier writers about protoplasm. Indeed, such distinguished biologists as the botanist NAEGELI and the zoologist OSCAR HERT-
WIG had felt it necessary to assume high values for protoplasmic viscosity
in order to make possible an interpretation of complicated life processes.
Thus in his Allgemeine Biologie (3rd edition, 1909), HERTWIG states: "Wie
NÄGELI und viele andere Forscher, sind wir der Überzeugung, daß die kom-
plizierten Erscheinungen des Lebensprozesses, vor allen Dingen der Ver-
erbung, nicht aus den Eigenschaften von Flüssigkeiten oder gelösten Stoffen
erklärbar sind." And he is of the opinion that the only thing that is fluid
about protoplasm is the water it contains. Strangely enough, some modern
authors still seem to think that protoplasm can never be truly fluid. Thus
SWANN, writing in 1956, states that "even the most fluid seeming protoplasm
is in fact a gel." This he says has been shown by various experiments [1].

[1] I have written to Dr. MICHAEL SWANN, asking what these experiments are,
but in his friendly and polite reply, he made no mention of them. Surely if
protoplasm is fluid and flows, by the very definition of sol and gel, it must be a sol.

Using his magnetic method, Heilbronn (1922) determined the viscosity of the protoplasm of various slime molds. In these studies because of the large size of the iron rods introduced into the protoplasm, measurements could be made only when the protoplasm was present in a rather large undifferentiated mass. Ordinarily the protoplasm of a slime mold may be in discrete strands, but at a time shortly before spore formation, the protoplasm may become aggregated into rather large masses (which later become subdivided by partitions). Heilbronn's values for the viscosity of the protoplasm of a number of slime molds are given in the following table.

Species	Viscosity in centipoises
Reticularia lycoperdon...............	16.5–8.5
Lycogale epidendron.................	15
Badhamia utricularis................	9.5–10.5
Physarium cinereum	9–11.5
Trichia fallax	11–16
Didymium serpula, race I	9–2.5
Didymium serpula, race II	11.5–12

The values shown in the table are for slime molds on glass. If they were placed on moist filter paper, the viscosity was only about two-thirds as great.

In interpreting the values that Heilbronn obtained for the protoplasmic viscosity both of the bean cells and of the slime molds, it should be remembered that the viscosity Heilbronn measured was not the viscosity of the hyaline fluid protoplasm alone, but the viscosity of the entire protoplasmic suspension, granules and all. In the case of the bean cells, it is apparent from an occasional remark of Heilbronn's that the starch grains he observed were falling through a typical protoplasmic suspension containing granules and other inclusions as well. And obviously when iron rods are twisted through a mass of slime mold protoplasm they encounter the resistance not only of the hyaline fluid protoplasm, but also the resistance due to the numerous granules and other inclusions which constitute a reasonably large fraction of slime mold protoplasm. Thus in comparing Heilbronn's values with the values obtained for other types of protoplasm, comparisons should not be made with values for hyaline (granule-free) protoplasm, but with values for the entire protoplasm, granules and all.

In 1928, Baas-Becking, Sande Bakhuyzen and Hotelling published a paper in which they record measurements made with a Brownian movement method in order to determine the viscosity of the protoplasm just under the cell wall of cells of the alga *Spirogyra*. This is a difficult paper to evaluate. According to the introductory statement, the second author was responsible for all of the observations, and one can but admire the patience and the care that went into what must have been a very trying series of measurements. The third author was the mathematician; but the senior author,

who presumably wrote the paper, was apparently not too well aware of
just what his collaborators had done. There are occasional contradictions,
not infrequent misquotations of the literature, and other evidences of
carelessness. The species of *Spirogyra* used in this study is not mentioned,
but presumably in the species used, as is generally true, the film of protoplasm which forms a cylindrical envelope about the rest of the cell is a
very thin film. Such a thin film is hardly ideal for Brownian movement
measurements. And as a matter of fact the authors (on page 11) state
that "parietal vacuoles, crystals, plastids and protoplasmic currents interfered materially." Added to this there must be another difficulty which
the authors do not mention, but which may well have been serious. In view
of the fact that the parietal protoplasm in *Spirogyra* lies between the outer
cylindrical cell wall and a second cylindrical membrane at only a small
distance from it, surely a granule in Brownian movement within this
narrow protoplasmic envelope of the central vacuole would more often
than not strike either the external boundary of the cell or the internal
boundary of the parietal protoplasm. In view of such a variety of
impediments to free movement, it is scarcely surprising that the observer
found a great variation in the distance which a granule would travel in
a given time. In the 176 sets of measurements, there is very little consistency, and calculation of the viscosity on the basis of the Einstein
formula gave widely divergent results. Indeed, if we exclude a single value
of 1 centipoise. no doubt due to an error, the other 175 measurements indicate viscosities which varied all the way from 2 centipoises to 308 centipoises. Various explanations are possible. The lower values, or perhaps
the lowest ones, may be the proper ones; for clearly if a granule did not
move very far in a given time, this may well have been due to the fact that
it had encountered one of the obstacles mentioned above. However there is
also another possible interpretation. The protoplasm of the *Spirogyra*
cell—like many other types of protoplasm—may consist of an outer rigid
layer or cortex within which lies more fluid protoplasm. Some of the
granules that Baas Becking and his associates studied may have been adjacent to the cortex or imbedded in it, and these granules would have much
less freedom of movement than those more nearly in the center of the
protoplasmic layer. All that one can be sure of is that some of the protoplasm in a *Spirogyra* cell is highly fluid.

A more satisfactory paper is that of Pekarek (1932). He used the
Brownian movement method he developed, a method described in an earlier
section. Pekarek states clearly that in order to obtain proper values of
the viscosity from Brownian movement studies, the protoplasm should not
be streaming or moving in any way, and that moreover the particles in
Brownian movement should not be at all close to each other. Accordingly,
he chose as the object of his studies the protoplasm in the terminal cells of
the rhizoids of the alga *Chara fragilis* Desv., for in these cells the protoplasm does not stream. At the very tips of the cells there is a mass of
rather hyaline protoplasm containing but few granules. Farther from the
tip, the protoplasm is much more densely granular. Pekarek found that

the viscosity of the hyaline protoplasm was 5 centipoises (at 22⁰ C.). Attempts to obtain values for the protoplasm rich in granules gave high values, which Pekarek is certain are incorrect, for the granules are too close to each other to permit reasonable freedom of movement. Accordingly, on the basis of calculations as to the concentration of suspended material in the granular protoplasmic suspension, Pekarek tried to calculate what the viscosity of the entire suspension would be. He used Einstein's formula for the viscosity of a suspension, and assuming a granular concentration of 10%, he obtained a value of 8 centipoises for the viscosity of the entire granular protoplasm. Another formula gave a slightly higher result, and for greater concentrations of granules, the calculated values for the viscosity were still higher. Pekarek concludes that the viscosity of the granular protoplasm, granules and all, is at least 8 centipoises, and possibly as much as 15 centipoises; the probable value he believes to be about 10 centipoises. Hence his results are in essential accord with the earlier work of Heilbronn on starch sheath cells and on the protoplasm of slime molds.

Thus, in so far as the evidence goes, and unfortunately it is not very plentiful, the indications are that the viscosity of the hyaline granule-free protoplasm of plant cells is of the order of 5 centipoises. The viscosity of the entire protoplasmic suspension, granules and all, may be several times as high as this or may be even higher, depending on the concentration of the granular material. This does not mean that all protoplasm in all plant cells has such a low viscosity. At the present time we scarcely know whether in most plant cells there is a cortex similar to that which is found in many different types of animal cells. Certainly in slime molds there is a thick cortex, but in the leaf and stem cells of ordinary plants, or in the cells of algae, the cortex may not be distinguishable from the outer membrane of the cell. At any rate it is with the interior protoplasm of a cell that we are primarily concerned, for the cortex will be discussed in another part of this treatise. Whether this interior protoplasm behaves like a Newtonian fluid is not definitely known, and it is possible that as the shearing force is varied, the viscosity may be different. However, in view of the fact that Heilbronn's earlier measurements were made by the gravity method, that is to say with a low shearing force, this is scarcely probable. Also in the Brownian movement studies of Pekarek the shearing force was negligible. Moreover in his studies of the huge slime mold, *Reticularia lycoperdon,* Heilbronn (1922) found that when he inserted iron rods into the interior protoplasm (so that they did not come near the outer cortex), the rods could be twisted over and over again without altering the viscosity. This is a clear indication that this type of protoplasm is not thixotropic and thus behaves like a Newtonian fluid.

But many botanists believe that protoplasm is elastic or that it shows the behavior of a non-Newtonian fluid. This opinion is to some extent based on the fact that a certain amount of force is necessary to dislodge the chloroplasts of *Spirogyra* and similar cells, and that when a greater shearing force is applied the apparent viscosity becomes progressively less (Northen

1936, 1938 a).   There is no doubt but that the outer membrane of a cell is
elastic and so also is the cortex.   Thus in ameba the cortical protoplasm
shows a thixotropic behavior (ANGERER 1936) and this is also true of the
cortex in the eggs of the worm *Chaetopterus* (WILSON and HEILBRUNN 1952).
In studies of the viscosity of protoplasmic droplets squeezed out of slime
molds and algae, PFEIFFER (1939, 1940) in work already referred to in the
section on methods, found a thixotropic behavior.   Actually the degree of
thixotropy was small.   In the case of protoplasm taken from the alga
Chara, the viscosity varied only from 4.98 to 4.08 centipoises when the
pressure of the shearing force varied from 7.12 cm. to 29.12 cm. of water.
As was pointed out previously, PFEIFFER was not dealing with living protoplasm, but with clotted and dead protoplasm, and it is surprising therefore
that a greater degree of thixotropy was not found.   Thus we can say with
some assurance that in so far as the valid evidence goes, the interior protoplasm of living plant cells is not thixotropic and has a low viscosity.

Protozoan Protoplasm

For only two types of protozoan protoplasm is there definite information concerning the absolute viscosity.   There have been measurements
of the viscosity of the protoplasm of ameba and there are also measurements of the viscosity of paramecium protoplasm.

In his well known and oft quoted description of the structure of
Amoeba proteus, MAST (1926) distinguishes between the inner flowing
protoplasm or plasmasol and the outer protoplasm which does not flow.
This outer region he calls the plasmagel.   However because a region of
protoplasm does not flow is not a necessary indication that it is solid, for
when fluid flows through a tube there may be a layer close to the walls
of the tube, the so-called inert layer, which does not flow.   Probably a part
of what MAST referred to as the plasmagel is really fluid.   However it is
easy to show that in *Amoeba proteus* low centrifugal force causes a movement of the granules in the interior of the cell but not the granules in the
outer layer.   This outer layer is strictly comparable to the cortex found in
cells generally and it is therefore advisable to refer to it as the cortex.
A similar cortex is found in the protoplasm of slime molds, which after
all are essentially giant amebae.   Thus in *Reticularia lycoperdon,* HEIL-
BRONN (1922) found the "ectoplasm" or cortex of this huge slime mold to be
several millimeters thick.   According to HEILBRONN, the viscosity of this
cortex is higher than that of the "endoplasm" or interior protoplasm and
may reach infinite values.

In various species of amebae, the thickness of the cortex varies widely.
In the giant ameba *Chaos chaos* (*Pelomyxa carolinensis*), the cortex seems
to be extremely thick, and the fluid streaming protoplasm may at times
occupy only a small fraction of the whole.   After these amebae are centrifuged so that they become rounded up, there may be several streams of
protoplasm separated by inert masses of protoplasm.   This condition is
similar to the ordinary condition found in slime molds, a condition in which

2*

there are numerous strands of protoplasm each surrounded by its own cortex. Because of the thickness of the cortex in *Chaos chaos* and because also the thickness of this cortex seems to vary widely, this form does not lend itself readily to measurements of protoplasmic viscosity. In measurements that have been made, one can scarcely be certain as to whether the information obtained pertains to the cortex, the interior protoplasm or both.

In *Amoeba dubia*, the cortex is so thin that it is somewhat difficult to recognize it. Thus it is easy to see through it and to make studies of the viscosity of the interior protoplasm. In *Amoeba dubia*, the cytoplasmic granules and inclusions are rather large and they fall readily through the protoplasm. Indeed they fall so readily under the influence of gravity that if the movement of the amebae is stopped by immersing them in dilute solutions of potassium chloride, the granules can be seen to fall within a few minutes[2]. And this in spite of the fact that the potassium ion, as will be shown later, causes a significant increase in the viscosity of the protoplasm. Hence in making viscosity tests with the centrifuge method, only very weak centrifugal forces are necessary.

According to Heilbrunn (1929 a), if specimens of *Amoeba dubia* are centrifuged with a force 128 times gravity at 18⁰ C., it takes approximately 3 seconds for the granules to travel through half of the cell, a distance of some 75 microns. This gives the velocity, the force is known, the average radius of the granules was found by measurement to be 2 microns. Hence to apply Stokes' law it is only necessary to know the specific gravity of the granules and the specific gravity of the fluid through which they move, along with the Cunningham correction discussed in the section on methods. The specific gravity of the amebae as a whole was determined by centrifuging them in various strengths of sugar solution; it was found to be 1.03. On the assumption that the hyaline protoplasm is essentially a protein solution with about 10% protein, the specific gravity of this protoplasm can be taken as 1.02. Because the total volume of the granules is about one-eighth of that of the entire ameba, it is possible to calculate the specific gravity of the granules; in this way a value of 1.10 is obtained. All that is now required is the Cunningham factor, q. The granules are on the average 11 microns apart; this gives a value for the Cunningham factor of 1.88. Using the corrected form of Stokes' law, the viscosity of the protoplasm of *Amoeba dubia* can then be calculated to be approximately 2 centipoises. This is the value at 18⁰ C. As will be shown later, the viscosity of the protoplasm of this ameba changes with temperature, and at 24⁰ C. the viscosity was found to be about 8/3 times as great, that is to say, approximately 5.3 centipoises.

It should be noted that this value for the viscosity of the interior protoplasm of *Amoeba dubia* was obtained on amebae that had rounded up as a result of the centrifugal treatment. However any uneven pressure such as

---

[2] Apparently such an effect does not occur if the amebae are raised in solutions relatively rich in calcium.

might be produced by the centrifuge would tend to raise the viscosity of
the protoplasm rather than to lower it. This follows from the work of
ANGERER (1936) to be discussed in a later section. That such low values for
the viscosity could be obtained with the centrifuge method is probably
due to the fact that only a few seconds were required for the tests. In the
case of the giant ameba, *Chaos chaos,* centrifugal treatment can not be used
for a study of the fluid interior protoplasm. But with a magnetic method,
ASHTON, in preliminary experiments done in HEILBRUNN's laboratory, was
able to obtain values which indicate a rather low viscosity for the flowing
protoplasm of the giant ameba (ASHTON 1957).

In 1930, working independently and using his Brownian movement
method, PEKAREK (1930 b) calculated the viscosity of the interior protoplasm
of an ameba at 24⁰ C. to be 5.2 centipoises. PEKAREK did not determine the
exact species of the ameba he studied. The closeness of the values obtained
by HEILBRUNN and PEKAREK is satisfying, but it must be admitted that the
extreme closeness is certainly coincidental. Different specimens of ameba
differ somewhat in their viscosity; moreover neither HEILBRUNN nor PEKAREK
claimed any great accuracy for their measurements . Indeed, PEKAREK
(1932) expressed concern over the fact that in his study of ameba he
stained the amebae with neutral red, and he thought it might be possible
that such staining would have an effect on the viscosity.

The important point is that the viscosity of the interior protoplasm of
ameba is certainly low; and not, as many had supposed, relatively high.
Thus PANTIN in 1924, on the basis of some rather uncertain deductions, had
concluded that the viscosity of the protoplasm of a marine ameba he used
in his experiments was over 1,000 centipoises. Indeed the idea of a very
fluid protoplasm seems to be repellent to some biologists, and from time
to time attempts are made to provide contrary evidence.

Thus, for example, HARVEY and MARSLAND (1932) in observing amebae
with the centrifuge microscope of HARVEY and LOOMIS (1930) state: "The
heavy crystals of *Amoeba dubia* always fall in 'jerks' even when moving
through a visibly clear field. They move and stop and move and stop as
if they met invisible obstructions. The same is true of *A. proteus.*" This
statement has been taken to mean that the interior protoplasm of an ameba
can not be fluid or at least can not be wholly fluid. In interpreting the
observation of HARVEY and MARSLAND, it should be recognized that observations with the HARVEY and LOOMIS microscope are not easy to make.
Certainly the photographs taken by HARVEY and MARSLAND are not impressive. And it is hard to understand that if a force of 128 times gravity can
move the granules or crystals of *Amoeba dubia* into half of the cell in
3 seconds, why there would still be granules or crystals in the interior
protoplasm after the amebae had been exposed to forces well over a
thousand times gravity for an appreciable time. MAST and DOYLE (1935),
who also observed amebae with the HARVEY and LOOMIS centrifuge microscope have a somewhat different story to tell. Thus in observing *Amoeba
proteus* with the centrifuge microscope, they state: "There were actual
streams of fat globules, and in these streams aggregations were now and

again formed in pockets, here and there, especially in small protuberances at the surface. However, the globules thus aggregated usually remained only a short time then broke away and rushed on again through the cytoplasm. Thus they continued with spasmodic interruptions until they reached the end of the ameba or the end of a pseudopod."

In *Amoeba proteus* it is not easy to make determinations of the viscosity of the interior protoplasm, for the rather rigid cortex is much thicker than the cortex in *Amoeba dubia* and it is hard to see through it. Looking through the cortex, it can be noted that many granular elements move rather readily under weak centrifugal forces. Something of this sort was observed by Mast and Doyle, although they made no effort to distinguish particles in the interior from those in the cortex. On the average, cytoplasmic granules and inclusions in *Amoeba proteus* are decidedly smaller than those in *A. dubia* and therefore, as might be expected from Stokes' law, it takes longer for these smaller inclusions to move under the influence of a given centrifugal force. If specimens of *Amoeba proteus* are centrifuged at low centrifugal speeds, it can without much difficulty be shown that the interior protoplasm of this ameba has a viscosity certainly not very different from that of the interior protoplasm of *A. dubia*. But it would be very difficult to make exact measurements, unless perhaps the amebae were sectioned.

As far as the present evidence goes, therefore, it seems safe to conclude that the protoplasm in the interior of an ameba under normal conditions is a fluid of relatively low viscosity. This is a fact of considerable importance, not only for our understanding of protoplasm generally; but also because if we are to draw any proper conclusions as to the mechanics of ameboid movement, it is essential for us to have correct information as to the physical nature of the moving protoplasm. It is for these reasons that all the evidence from various sources was thought worthy of consideration.

The protoplasm of paramecium is quite different from that of ameba. Inside of a rigid cortex, there is a very dense suspension of granular material. So concentrated is this suspension that even after prolonged and vigorous centrifugation the granules are unable to move any appreciable distance. The conclusion to be drawn is that the concentration of suspended material in the paramecium protoplasm must approach the limiting value of 74.05%. This is very different from the situation in *Amoeba dubia*, in which according to Heilbrunn (1929 a), the granular concentration is only about 12%.

Although it is not possible to study the viscosity of the hyaline protoplasm of paramecium—that is to say the viscosity of the suspension medium—it is possible to study the viscosity of the suspension as a whole. This can be done by having the animal ingest particles of one sort or another; for example, finely divided starch, or iron or carmine particles. The food vacuoles containing these ingested particles can then be moved through the protoplasm by centrifugal force. The first attempt to measure the viscosity of the protoplasmic suspension in paramecium was made by

FETTER (1926). She fed paramecia powdered starch and powdered iron and determined how long it took for the vacuoles containing these materials to move through the protoplasm of a paramecium when the animals were centrifuged with a force 709 times gravity. The starch moved through the cells in 2½ minutes, the iron in 3½ minutes. After determining the specific gravity of moistened starch powder and moistened iron powder and considering the specific gravity of the paramecium protoplasm as equal to that of the paramecium as a whole, she was able to apply STOKES' law and calculate the viscosity. Her data for starch particles indicate a value of approximately 8,000 centipoises for the viscosity; with the iron particles she obtained a value of 8,700 centipoises. Apparently FETTER's values for the specific gravity of the starch and iron are not very accurate, and indeed she makes no allowance for the fact that the starch and iron are contained in vacuoles. Nevertheless there is no doubt that for the centrifugal force she used, the viscosity of the paramecium protoplasm is of the order of thousands of centipoises.

Indeed the FETTER experiment has for years been done as a class exercise in the course in general physiology at the University of Pennsylvania, and values of thousands of centipoises have regularly been obtained. In this course exercise, care is taken to make proper estimates of the specific gravity of the food vacuoles, for it is a simple matter to count the number of starch or iron or carmine particles in the vacuoles and, knowing the specific gravity of these particles, to make reasonable deductions as to the average specific gravity of the vacuoles. However, BROWN (1940) repeated the FETTER experiment and instead of finding high values for the viscosity of the protoplasmic suspension concluded that the viscosity was only 50 centipoises or less.

Unfortunately, neither FETTER nor BROWN mentioned the species of paramecium that was used in the experiments, and it might be thought that the difference in viscosity values was due to differences in the organism used. However, there is a much better explanation.

HEILBRUNN (1928) wrote "It is very probable that the viscosity of the paramecium protoplasm varies with different rates of shear. It would be interesting to make determinations of absolute viscosity at different centrifugal speeds." Not until very recently was this done. In a doctoral thesis written in 1955 at the University of Pennsylvania, LARSON presents the results of careful and efficient measurements on the viscosity of various species of paramecium. Microfilm copies of this thesis can be obtained from the University of Pennsylvania; a somewhat shortened version of the thesis is now in preparation.

LARSON made measurements on four species of paramecium. Using the same type of procedure as that previously employed by FETTER, but with more accurate determinations of the specific gravity of the moving food vacuoles, LARSON found for *P. caudatum*, when the animals were allowed to ingest powdered carmine and the centrifugal force was varied from 213 times gravity to 3.144 times gravity, that the viscosity as determined by STOKES' law varied from 510 centipoises to 7,610 centipoises. Actually as the centrifugal force was increased the velocity of movement of the food vacuoles stayed constant. Clearly, the protoplasmic suspension in *P. caudatum* is dilatant. Larson's estimates of the specific gravity of the food vacuoles were apparently accurate, for she obtained concordant values for the viscosity when the paramecia were fed starch grains or finely powdered iron instead of carmine. The values for the starch-fed paramecia showed a viscosity variation from 440 centipoises for a force 171 times gravity to 7,760 centipoises for a force 3,144 times gravity. With finely powdered iron, the viscosity was only 165 centipoises at 25 times gravity and

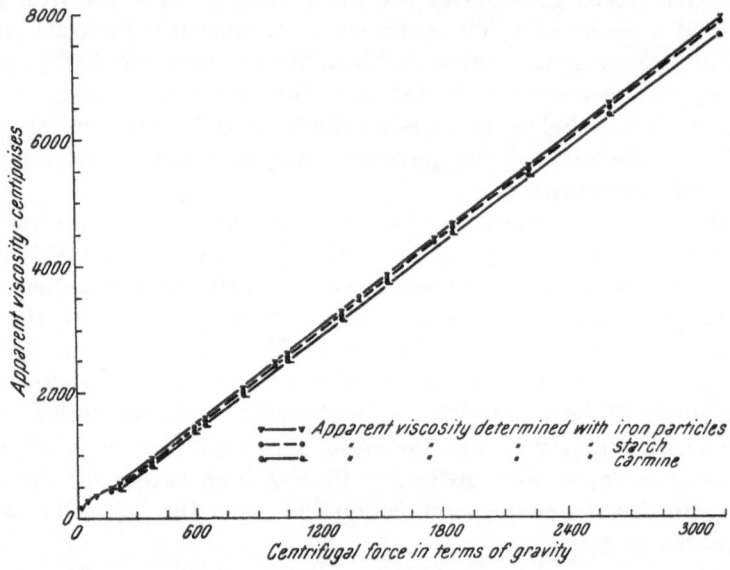

Fig. 1. Variation of the viscosity of the protoplasm of *Paramecium caudatum* with changes in shearing force. (Larson.)

7,780 centipoises at 3,144 times gravity. All these results are shown graphically in Fig. 1. With forces less than 25 times gravity, the *Paramecia* do not orient well in the centrifuge tubes and the moving food vacuoles may come to a stop at the side rather than at one end of the animal.

Larson's results show clearly that the protoplasm of *P. caudatum* is dilatant, and they also serve to explain the divergent findings of Fetter and Brown. Actually her values were never as low as Brown's nor as high as Fetter's for similar centrifugal forces. This, according to Larson, is due to the fact that both of these authors made errors in their calculations. Thus from Brown's measurements, Larson calculated the viscosity to be 94 rather than 50 centipoises, and in Fetter's measurements, there were errors in the estimates of specific gravity.

The fact that the protoplasm of *P. caudatum* is dilatant is a clear indication that the protoplasm in which the granular material is suspended is a fluid of low viscosity. For according to Röder (1939), dilatancy is only possible if the liquid in which the particles are suspended has a low viscosity.

Larson found also that the protoplasm of *Paramecium multimicronucleatum* showed the same sort of dilatancy as did that of *P. caudatum.* With forces 197 to 3,144 times gravity, the viscosity varied from 490 to 7,910 centipoises.

In two smaller species of paramecium, *P. aurelia* and *P. trichium,* the protoplasm showed dilatant behavoir as the centrifugal force increased up to 1,000 times gravity, but with still higher forces the protoplasm became thixotropic. This is shown in Fig. 2. The fact that a suspension can be both dilatant and thixotropic for different ranges of centrifugal force is an interesting phenomenon and one worthy of further study.

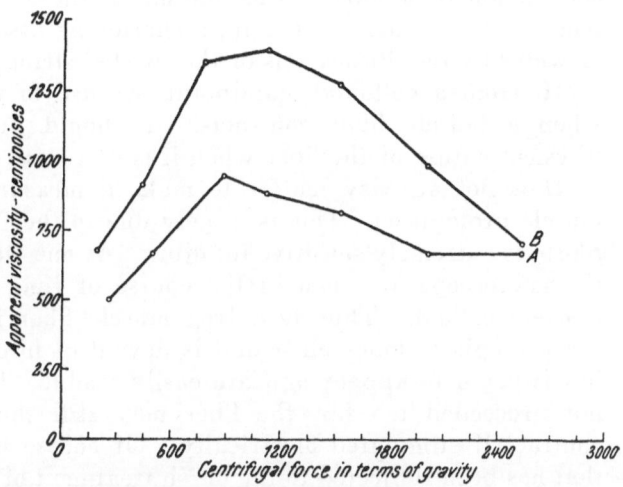

Fig. 2. Changes in the viscosity of the protoplasm of two relatively small species of paramecium as the centrifugal force is varied. Curve *A* shows data for *P. trichium*, curve *B* for *P. aurelia.*
(Larson.)

The conclusions to be drawn from the experimental evidence are now clear. No doubt because of the fact that the protoplasm of paramecium is a highly concentrated suspension, the viscosity of the entire protoplasm—granules and all—shows the behavior of a non-Newtonian liquid, and just as is the case of various inanimate suspensions, in which the suspended material is present in high concentration, the paramecium protoplasm is dilatant. Moreover, it seems certain that the non-granular or hyaline protoplasm of paramecium has a low viscosity, for this is a characteristic of the suspension medium of dilatant systems. Such a low viscosity of the non-granular protoplasm could almost be deduced from the fact that in the paramecium the food vacuoles move through the protoplasm without apparently meeting with much resistance as they follow a rotary course through the cell. Thus it seems evident that both in ameba and in paramecium the hyaline protoplasm has a low viscosity, and for the present at least we may conclude that this is a general characteristic of the protoplasm in the interior of protozoan cells.

Muscle Protoplasm

Although, over a long period of years, many physiologists have thought and have written about the viscosity of muscle protoplasm, there has been little in the way of exact measurement. In 1863, the great physiologist Kühne happened to see a nematode worm within the protoplasm of a frog muscle fiber. The worm swam back and forth along the length of the fiber and this led Kühne to conclude that the interior of the fiber was fluid. Such a conclusion was not necessarily valid, at least for the normal fiber, for the presence of the worm might well have changed the physical state of the protoplasm. And so students of the physiology of muscle continued to argue the question as to whether the interior of the muscle was a fluid sol or a more or less rigid gel. Some of those who sought to explain the contraction of a muscle in physical terms believed that when a muscle fiber shortened, its protoplasm changed from sol to gel; others held quite the reverse opinion, and indeed for many years it was commonly believed that the contraction of a muscle was similar to the shortening of a violin string when it was immersed in acid, a shortening associated with the taking up of water by the fibrous gels of the twisted string.

If, from a colloidal standpoint, we are to understand what happens when a muscle fiber contracts, we should know something about the physical nature of the fiber when it is in a state of rest.

It is not an easy matter to make a measurement of the viscosity of muscle protoplasm. This is largely due to the fact that the muscle protoplasm is extremely sensitive to injury. As one watches a muscle fiber under the microscope, one can in the course of time see changes in the protoplasmic colloid. Thus if a frog muscle fiber is carefully dissected out, the protoplasm looks clear and is devoid of fibrils. But before very long, fibrils begin to appear and are easily visible. If this fibril formation has not proceeded too far, the fiber may still show signs of life and will contract if stimulated electrically. Of course in an injured fiber, or one that has been subjected to the harsh treatment of fixation by one or another of the reagents used by cytologists, fibrils are very evident. Beyond much doubt in the living uninjured fiber there are proteins present which have elongated molecules and these proteins are presumably the ones which form the fibrils visible either with an ordinary microscope or with an electron microscope. This fibril formation is but one indication of the sensitivity of the protoplasmic colloid of muscle fibers. Another indication is that the surface precipitation reaction in muscle protoplasm is rapid and apparently involves extensive clotting of the protoplasm.

If one is to make determinations of the viscosity of the protoplasm in the interior of a muscle fiber, it might be thought that the centrifuge method could be used. Although cytoplasmic granules are not ordinarily visible in muscle fibers, in the muscle fibers of frogs, nuclei are present throughout the protoplasm, and it might be thought that these could be moved by centrifugal force. But when frog muscle fibers are centrifuged, the nuclei do not move, at least they did not move in a few hastily executed tests. This lack of movement may be due to injury produced in the very sensitive protoplasm. Perhaps if intact muscles were centrifuged for long times and were then sectioned and studied carefully, it might be possible to detect some shifting of nuclei, but this has never been done.

Up until the present time, there has been only one set of measurements made on the viscosity of muscle protoplasm. Rieser (1949 a) injected small oil drops into isolated muscle fibers of the frog, and he then placed these fibers on the stage of a microscope held in a vertical position so that he could observe the speed with which the droplets rose through the protoplasm of the muscle fiber under the influence of gravity. Then, knowing the specific gravity of the oil, it was only necessary to know the diameter of the drop, its speed of movement and the specific gravity of the protoplasm through which the drop moved in order to apply STOKES' law for a measure of the viscosity. RIESER measured the specific gravity of the muscle fibers as a whole by centrifuging them in various concentrations of sugar solution of known specific gravity, and he assumed that the specific gravity of the protoplasm was not essentially different from that of the entire fiber. This all sounds very simple, but as a matter of fact it is a very difficult matter to introduce an oil drop into the interior of a muscle fiber with a minimum of injury. In spite of the fact that RIESER is an accomplished microdissectionist and is very skillful in such manipulations, it was only rarely possible for him to accomplish the injection of the oil drop without injuring the fiber to such an extent that the drop could not move. Also when the drop was put in at an angle, as it usually was, it could only move away from the direction from which the micropipette was inserted. And once the oil had moved through the fiber, the protoplasm in its path was sufficiently altered so that when the fiber was turned 180⁰, the oil drop did not move back again. By dint of long patient effort, and after many unsuccessful attempts, RIESER was finally able to observe 10 cases in which the introduction of the oil drop produced so little injury that the drop was able to move under the influence of gravity. The measurements made as a result of these 10 successful experiments are given in the following table.

Viscosity of the protoplasm of isolated frog muscle fibers

Fiber no.	Oil used	Diameter of oil drop, μ	Diameter of fiber, μ	Velocity μ/sec.	Viscosity centipoises
1	Paraffin oil	16	90	1.400	14
2	,, ,,	21	121	1.470	15
3	,, ,,	19	96	0.167	103
4	,, ,,	20	108	0.620	37
5	Oleic acid	15	117	0.670	22
6	,, ,,	17	143	0.433	45
7	Butyl stearate	24	105	0.133	280
8	,, ,,	32	114	0.433	145
9	,, ,,	22	105	0.733	45
10	,, ,,	11	124	0.433	24

In calculating the viscosity in the above table, because of the relatively large size of the oil drop in relation to the diameter of the fiber, RIESER was obliged to use the LADENBURG correction to STOKES' law (see LADENBURG 1907). The oil drops introduced by RIESER were never spherical within the fibers. Rather they were slightly ovoid, the ratio of large to small axis in all cases being approximately 3 to 2. Probably this ovoid shape is due to the fibrous nature of the proteins of the muscle fiber. In interpreting the viscosity values he obtained, Rieser emphasizes the fact that the lower values are undoubtedly more nearly the correct ones, for even the slightest injury causes a retardation or stoppage of movement and it is not possible to introduce a micropipette into a muscle fiber without some injury. Rieser noted also that the oil drops never moved through the protoplasm "of moribund fibers in which longitudinal fibrillar elements had become more or less distinct."

An attempt to repeat Rieser's measurements has recently been made by Hiramoto (1956). However, Hiramoto never saw any movement of the injected oil drops. This is readily understandable, for Rieser had to work hard and long before he was able to introduce oil drops with so little injury to the protoplasm that the oil drops were able to move through it under the influence of gravity. Moreover, from Hiramoto's photographs it is obvious that the fibers he studied had visible fibrils in their protoplasm; hence they were probably not in as good condition as they might have been. Doubtless the fluid protoplasm becomes more and more restricted as the fibrils form, although there may still be enough of the fluid left so that contraction can occur. In Hiramoto's photographs the oil drops within the fibers are much longer and narrower than those described by Rieser; this could be due to the presence of fibrils.

That the interior protoplasm of a muscle fiber is really a fluid and not a gel as so many have claimed is indicated by the fact that when aqueous solutions are injected into a fiber these solutions spread through the protoplasm of the fiber freely. This is especially well shown when solutions containing a vital dye such as methylene blue are injected. If the protoplasm were a solid gel, then a small amount of an injected solution would remain as a discrete droplet. This is actually what happens when aqueous solutions are injected into a muscle fiber which had been previously exposed to glycerol so that most of its water had been removed.

In addition to determining the viscosity of the interior protoplasm of frog muscle fibers, Rieser also made another observation of importance. In his microinjection experiments, he sometimes injected oil too rapidly or in too large amounts. If large amounts of oil, or of air, were injected, they never occupied the entire width of the fibers. Rieser states that "Always there was a region directly under the sarcolemma which the oil or air could never displace." This outer region represents the cortex of the muscle cell—in the frog muscle fiber it has a thickness of approximately 10 microns. Strangely enough, in fibers in which the entire interior was displaced by oil, the cross striations appeared more plainly than in fibers with their interior protoplasm intact. This might give rise to the suspicion that the cross striations in the healthy living fiber exist only in the cortex. But how then could one explain the fact that when sections of muscle fibers are examined either under the light or the electron microscope, striations are clearly visible in the interior of the muscle protoplasm? Could it perhaps be possible that it is only after treatment with the coagulative agents used in the fixation that precedes sectioning that both fibrils and striations appear as artefacts through the fiber? Certainly if the striations of muscle fibers represent, as commonly supposed, partitions which run clear through the fibers, then it would be difficult to understand how it is that a droplet of oil, merely under the influence of gravity, can pass up through the center of the muscle fiber without disturbing any of the striations.

For the present, on the basis of RIESER's study, we may conclude that a striated muscle fiber consists of an outer stiff cortex and an inner mass of protoplasm which is fluid. In all probability, this inner protoplasm is even more fluid than RIESER's measurements indicate, for even in his best experiments there must have been some injury with consequent increase in viscosity.

Nerve Protoplasm

Our knowledge of the physical properties of nerve protoplasm is even more fragmentary than our knowledge of the physical properties of muscle protoplasm. Like the protoplasm of muscle, the protoplasm found in nerve fibers is undoubtedly sensitive to injury; and so in interpreting the literature, one must be careful to remember that what is true of injured protoplasm is not necessarily true of normal protoplasm.

When a nerve fiber or a nerve cell is cut, the protoplasm flows out. At any rate, such an outpouring of nerve protoplasm has often been noted by various observers. Thus HAECKEL (1857), YUNG (1878), and KRIEGER (1880) all saw the protoplasm flow out of the cut nerve fibers of crustacea; and even before them, HELMHOLTZ (1842) commented on the fluidity of the protoplasm of crustacean nerve fibers. Later HARDY (1895) as a result of his observations on crayfish nerves became convinced that their protoplasm was highly fluid. When air bubbles were present in these nerves, the bubbles moved about freely. FLAIG (1947) saw protoplasm flow from the cut end of the giant axon of the squid. SMITTEN (1946) observed nerve cells in an autonomic ganglion of the frog heart. When these cells were pierced with a microneedle, the protoplasm flowed out.

No one of these observations is at all quantitative, and all of them must be judged in the light of the fact that when a cell is torn or broken, a clotting reaction sets in, a reaction which HEILBRUNN (1927) has called the surface precipitation reaction. That such a reaction does occur in the protoplasm of nerve fibers has long been known. Thus HAECKEL (1857) states that after a crustacean nerve fiber has been cut, the protoplasm that flows out clots "in the form of droplets, threads and granules." A beautiful surface precipitation reaction has been described for the protoplasm of the giant nerve axon of *Sepia officinalis* by YOUNG (1936). Indeed any injury to this or other types of nerve protoplasm may convert the protoplasm into a fibrillar mass of high viscosity. According to FLAIG (1947), such fibrillation occurs when a giant nerve axon of the squid is punctured. In preparing these axons for study, they are frequently injured severely, so that this injured protoplasm may act as though it had a very high viscosity. If the squid giant nerve fibers are examined *in situ,* they appear to have a fluid interior surrounded by a more solid cortex. At any rate such an conclusion can be drawn from the observations of Chambers and Kao (1951). These investigators injected solutions of acid dyes into the interior of the fiber, and found "definite evidence of a fluid core. When micro-injected, the solution takes a preferred route of running along the central core through the nerve. The dye then diffuses radially."

The only quantitative measurement of the viscosity of nerve protoplasm is due to Rieser (1949 b). In a small note, he mentions the fact that he injected oil droplets into the protoplasm of the large nerve fibers of the ventral nerve cord of the lobster and watched these droplets rise under the influence of gravity. The viscosity value he obtained from these measurements was 3 centipoises. It should be noted that when oil drops are injected into the lobster nerve protoplasm they assume a spherical shape and are not ovoid as are the oil drops injected into muscle protoplasm.

The Protoplasm of Marine Eggs

For the student of the physical properties of protoplasm, the eggs of marine invertebrates constitute ideal material, and many studies have been made of the protoplasm of the eggs of sea urchins, starfish, worms and clams. These isolated cells, easily obtainable at certain seasons of the year, are of a convenient, constant size. Usually, the viscosity of the protoplasm of these eggs can readily be studied by the centrifuge method, although in some instances the protoplasm is so full of granules that, as in the case of paramecium, no shift of granules is possible. Some types of eggs can conveniently be obtained in an immature condition, and such eggs have a large nucleus, commonly called the germinal vesicle. The germinal vesicle has a single large nucleolus. In some, but not all of these immature eggs, the nucleolus falls through the nucleus under the influence of gravity and this makes possible a determination of the viscosity of the nuclear sap.

The eggs of various species of sea urchins have been widely used, but perhaps none as often as those of the Woods Hole sea urchin, *Arbacia punctulata.* Indeed the egg of *Arbacia punctulata* (Lamarck) has become a standard object of study. The information concerning this egg is voluminous and many-sided. Fortunately Dr. Ethel B. Harvey (1956) has collected all the available data in a volume which serves as a valuable guide for anyone interested in working with sea urchin eggs.

Heilbrunn (1926 b) measured the absolute viscosity of the protoplasm of the *Arbacia* egg. This constitutes the first quantitative measurement of the viscosity of the protoplasm of any animal cell, and it is also the first measurement of the viscosity of the protoplasmic fluid in which granules or other inclusions may be suspended. As the work on paramecium has indicated, the viscosity of the entire protoplasmic suspension may be highly variable depending on the concentration of the suspended material. But as far as we know now, it seems possible that in many diverse types of cells, the viscosity of the protoplasm itself, or rather the viscosity of

the fluid medium in which granules and inclusions are suspended, is rather
uniformly low if we consider only the protoplasm of the cell interior when
this is in the resting state.

To centrifuge *Arbacia* egg cells and to estimate the speed at which the
cytoplasmic granules move through the protoplasm is rather a simple
matter. With a force 4,968 times gravity HEILBRUNN found that it required
approximately 25 seconds for the granules heavier than the protoplasmic
fluid to move into the centrifugal two-thirds of the egg, and the great
majority of the granules have moved into half of the egg in 40 seconds.
When nearly all of the granules have been shifted into half of the egg,
it can be assumed that the granules originally at the pole farthest distant
from this half are now at the equator. Thus, since the diameter of the egg
is 73 microns, the speed of the granules for a force 4,968 times gravity is
apparently 0.00009 cm. per second. However, when the eggs are centrifuged they elongate and under the conditions of the experiment they may
become 85 to 90 microns long. Thus a correction is neccessary, and a more
exact value for the granular speed is 0.00011 cm. per second.

In order to apply STOKES' law it is necessary to know the specific gravity
of the granules and also the specific gravity of the fluid through which they
travel. By breaking up the eggs and centrifuging the homogenated material
in various strengths of sugar solution, HEILBRUNN arrived at a value of
approximately 1.11 for the specific gravity of the granules. (This determination was made at a time before it was known that the presence of
calcium in sea water caused the protoplasm to clot when the cells were
broken. Hence the presence of clotted strands interfered with the
measurement. It would have been better to break up the eggs in calcium-
free sea water.) The specific gravity of the entire eggs can also be
determined by centrifuging in various strengths of sugar solution; and
then, knowing the concentration of granules in the egg, by a simple
arithmetical calculation it is possible to arrive at a value for the specific
gravity of the non-granular protoplasm. The value that HEILBRUNN took for
the difference in specific gravity between granules and the fluid through
which they move was 0.072.

Two other values needed to be determined. It is necessary to know the
radius of the granules, and also the CUNNINGHAM factor. The radius can be
determined by direct measurement, but in view of the fact that measurements of very tiny particles can not be made accurately, the values that
can be obtained are far from exact. HEILBRUNN took as the value for the
radius of the granules 0.000016 cm. From a knowledge of the radius of the
granules and the distance that separates them, a computation of CUNNING-
HAM's factor is possible. The value arrived at was 10.54.

Substituting all these values into the equation for STOKES' law
(with CUNNINGHAM correction), the viscosity of the granule-free protoplasm
of the *Arbacia* egg was found to be 1.8 centipoises. This value is based only
on a measurement of the speed of movement of the colorless (yolk) granules
which constitute the major portion of the granular inclusions of the *Arbacia*
egg. But there are also relatively large pigment granules and these also

move under the influence of centrifugal force. They move more rapidly
than the smaller granules, doubtless because of their larger size, but their
movement is impeded by the fact that they are continually colliding with
the smaller granules.

With a force 4,968 times gravity, the speed of the pigment granules
through the egg was found to be about 0.00024 cm. per second. On the
assumption that their specific gravity is not essentially different from that
of the smaller colorless granules, Heilbrunn obtained a value of 6.9 centipoises for the viscosity of the fluid through which they pass. But this
value takes no account of the fact that the movement of the larger
pigment granules is constantly impeded by their collisions with the
slower moving smaller granules. It is possible to calculate the decrease
in speed of the larger granules as a result of their collisions with the
smaller ones. On the assumption that the granules are inelastic,
that they are smooth enough so that no rotation follows a collision,
and also that when the particles collide, the line joining their
centers is along their path of motion, Heilbrunn was able to determine the effect of the collisions in slowing their speed. As a matter of
fact, no one of these assumptions is strictly correct, but we are interested
only in an approximate solution. The calculation, details of which are
given in Heilbrunn's paper, shows that the effect of the collisions is to
reduce the velocity of the larger granules by a factor of approximately 2.8.
Therefore, dividing 6.9, the uncorrected value for the viscosity by 2.8, we
can obtain a value for the viscosity of the granule-free protoplasm. This
is 2.5 centipoises, a value not very different from the value of 1.8 centipoises determined from the rate of movement of the smaller granules.

As a matter of fact, the original uncorrected value for the viscosity of
the medium through which the larger granules travel is really a value for
the viscosity of the protoplasmic suspension consisting of the hyaline
protoplasm plus the smaller granules. If we were to apply various for-
mulas for the viscosity of a suspension in terms of the viscosity of the
suspension medium and the concentration of suspended material (see
section on the Viscosity Concept), we would obtain values for the viscosity of the entire protoplasmic suspension that would be two or three
times that of the viscosity of the hyaline protoplasm. Thus the value of
6.9 centipoises obtained for the viscosity of the entire protoplasm is more
or less what it should be if the viscosity of the hyaline protoplasm is taken
at 2 centipoises.

It should be clearly recognized that the values for the viscosity obtained
by Heilbrunn are not in any sense exact values. There are various possi-
bilities for error. In the original calculation of the specific gravity of
granules and the medium in which they are suspended, Heilbrunn took the
concentration of granular material as approximately 20%. According to
Costello (1939), it is decidedly more than that. From Costello's figures,
correcting for the spaces between the granules when they are closely
packed, the concentration of granular material may be as much as 40%.
If this is true, it would alter the value calculated for the specific gravity

of the fluid surrounding the granules and this value would be somewhat less. A recalculation on the basis of the higher value for the concentration of granular material would give a viscosity value for the hyaline proto-plasm of 2.04 instead of 1.8 centipoises. However, the CUNNINGHAM factor q was also calculated on the basis of a 20% concentration of granular ma-terial, and thus the true value for q would be larger than the one that was used. This would no doubt compensate, or more than compensate, for the error in the determination of the specific gravity of the hyaline protoplasm. But a more serious error would be involved if the radius of the granules was not accurately measured, for this quantity is squared in STOKES' law. This difficulty has often been emphasized by HEILBRUNN (1926 b, 1928, 1952, 1956). HARVEY (1932 a) believes that the colorless granules of the *Arbacia* egg are decidedly larger than they were estimated to be by HEILBRUNN.

Fortunately, there is another method for determining the viscosity of the *Arbacia* protoplasm. After the eggs are centrifuged so that the heavier granules are all in one half of the egg, the granules return by Brownian movement and after a time they are again scattered through the egg. If we could estimate the time required for these granules to return, we could then calculate the viscosity of the fluid through which they move on the basis of the EINSTEIN formula for Brownian movement (see section on The Measurement of Protoplasmic Viscosity). This was done by HEILBRUNN (1928). Although no exact measurement is possible for the time it takes for the granules to move from the equator to the light pole, the time is less than an hour and more than a half hour. If we take this time to be 45 minutes, then the viscosity of the protoplasm through which the granules pass is found to be 5 centipoises [3]. Actually the EINSTEIN formula is not an exact expression and it certainly can not be expected to give accurate values when the concentration of suspended particles is high. One ad-vantage of the Brownian movement method is that it requires no deter-mination of the specific gravity either of the particles or of the medium through which they travel. However, knowledge of the radius of the moving particle is important, and as has been pointed out, exact measure-ment of the small colorless granules of the *Arbacia* egg is not possible. An error in this measurement would act in opposite fashion for the two types of viscosity determination. For if the granules were really larger than we have taken them to be, this would make the values obtained with the centri-fuge method too low; and on the other hand, it would make the values obtained with the Brownian movement method too high. Thus it seems wise to average the two values and to assume that the viscosity of the

[3] In 1928, HEILBRUNN arrived at a value of 4 centipoises. In his calculation he used an EINSTEIN equation in which a low value was taken for the AVOGADRO constant. At present the usual value is 6.03×10^{23}, but in the past values of from 4 to 7 times 10^{23} have reported. It may be noted that values for physical constants are not always perfect when first obtained, so that physicists and physically-minded biologists should not be too critical of the uncertain values sometimes obtained by biologists.

hyaline protoplasm of the *Arbacia* egg is approximately 3.5 centi-poises.

Sea urchin eggs vary widely in the concentration of suspended material in the protoplasm. So, for example, the protoplasm of the *Paracentrotus lividus* egg is so full of granular material that this can scarcely move at all when the egg is centrifuged. Hence this protoplasm may well behave like the protoplasm of paramecium (see earlier section). According to RUNNSTRÖM (1928 a, 1928 b), the protoplasm of the egg of the sea urchin *Echinocardium cordatum* contains in addition to small granules, larger "Flocken" or flaky precipitates, and vacuoles. In order to use the centrifuge method for the determination of the protoplasmic viscosity of the eggs of this species, one should know the total concentration of granular material in order to apply the CUNNINGHAM correction. Also one should be very careful that the protoplasm one is studying is normal and not moribund. The sea water in the Naples Stazione Zoologica is often rather foul—or at any rate it has been in the past—and it is understandable why workers at this laboratory have sometimes published papers on sea urchin eggs which did not show anything like uniform behavior following fertilization. At Woods Hole, workers in the Marine Biological Laboratory have learned to discard eggs that for one reason or another do not show over 90% cleavage on fertilization. Even eggs that are in bad condition, for example as a result of aging, can show fairly high percentages of cleavage, for some of the protoplasm they contain is still normal.

Anyone who has ever centrifuged marine eggs which have a relatively low concentration of granular material can not fail to be impressed with the ease with which the granules move through the protoplasm. So, for example, if the eggs of the worm *Chaetopterus pergamentaceus* are centri-fuged with a force of approximately 2,000 times gravity, in a matter of just a few seconds, the granules move much of the distance through the egg. Undoubtedly the movement is facilitated by the fact that relatively few granules are present and also by the fact that in the *Chaetopterus* egg the number of lighter granules, presumably fat particles, is greater than it is in the *Arbacia* egg. Thus when the heavier granules move toward the centrifugal end of the egg the movement of a relatively large number of particles in the opposite direction facilitates the return flow. In eggs of this type, the CUNNINGHAM factor becomes less important and it may well be neglected; although if it is neglected, one can only obtain values for the viscosity which are maximum values.

No one has ever attempted a measure of the absolute viscosity of the protoplasm of the *Chaetopterus* egg, but HEILBRUNN (1926 b), using the same methods that he used for the *Arbacia* egg, but neglecting the CUNNINGHAM factor, did determine the viscosity of the granule-free protoplasm of the egg of the clam *Cumingia tellinoides*. This beautiful egg, unfortunately no longer available at Woods Hole, has protoplasm so fluid that when it is centrifuged for only four seconds with a force 4,968 times gravity, most of the heavy granules move into one half of the egg, although some of them are trapped by the mitotic spindle and asters of the first maturation division. Values for the specific gravity of granules and hyaline protoplasm are easy to obtain. The value for the viscosity of the (granule-free) protoplasm is then found to be 4 centipoises; and this is a maximum value, for the CUNNINGHAM factor was neglected.

HARRIS (1935) attempted to determine the viscosity of the granule-free cytoplasm of the egg of the worm *Sabellaria alveolata* by studying the rate of Brownian movement in eggs which had previously been centrifuged. He made motion pictures of granules in the hyaline zone and then applied the EINSTEIN formula for Brownian movement in order to obtain a value for the viscosity of the fluid in which the granules were moving. There was a disturbing factor, which according to HARRIS, introduced a serious complication. The granules did not seem to move in random fashion, but their movement seemed to be guided in one direction or another. Hence HARRIS speaks of streaming movements which interfered with his measurements. Accordingly he made a correction for what he thought to be streaming movements, and as a result of this correction he obtained a value of 20 centipoises for the viscosity of the hyaline protoplasm. If no correction had been made, the viscosity value would have been some small fraction of 20 centipoises, for as in the *Arbacia* egg the granules showed a complete return to their original position in about an hour. HARRIS is at a loss to explain the streaming movements he observed, and he suggests the possibility of an influence exerted by gravitational force or by osmotic phenomena; he also refers to previously observed movements of protoplasm during the course of cell division. But the *Sabellaria* egg that HARRIS studied, as he himself points out, is in a quiescent stage in the metaphase of its first maturation division. This is exactly the same stage the *Cumingia* egg is in some short time after it is shed into sea water. Accordingly, in the hyaline zone of the centrifuged *Sabellaria* egg, just as in the hyaline zone of the centrifuged *Cumingia* egg, there presumably is a mitotic spindle with its asters and these would interfere with the free movement of granules. Hence it is readily understandable that the granules might not move freely but would directed in their movement by the presence of these obstructions. There is also a possibility that the free movement of granules is influenced by their electric charges and the relation of these charges to the charges on the outer surface of the egg and on the chromosomes present on the spindle.

From the evidence accumulated thus far, it is apparent that the protoplasm of some marine eggs, at least at certain times, has a rather low viscosity, a viscosity such as would be expected for a protein solution. The question now arises as to whether this fluid protoplasm behaves like a Newtonian fluid. If it were non-Newtonian then it could be argued that the high shearing force used in the determination of viscosity by the centrifugal method might be responsible for what to many biologists has seemed to be a surprisingly low value for protoplasmic viscosity.

HEILBRUNN (1926 a) determined the effect of varying the centrifugal force on the values obtained for the viscosity of *Cumingia* egg protoplasm, and the data were recalculated by Heilbrunn (1928). The results are shown in the following table.

Centrifugal force in terms of gravity	Viscosity of *Cumingia* egg protoplasm in arbitrary units
310. 9	3.5
552	3.7
1242	3.5
4968	3

In the table, although the viscosity is given in arbitrary units, these units happen to be close to the actual values of the viscosity in centipoises. It is clear that the viscosity remains constant from 310.5 times gravity to 1,242 times gravity. The value for a centrifugal force of 4,968 times gravity is not very accurate and presumably is too low. For at this relatively high centrifugal force, the granules move so rapidly that the time required for their movement is only a matter of 3 seconds. It is difficult to stop the centrifuge at the exact expiration of 3 seconds and the centrifuge tubes turned for a fraction of a second before they could be stopped. Thus the true value for the viscosity is a little higher than the one recorded.

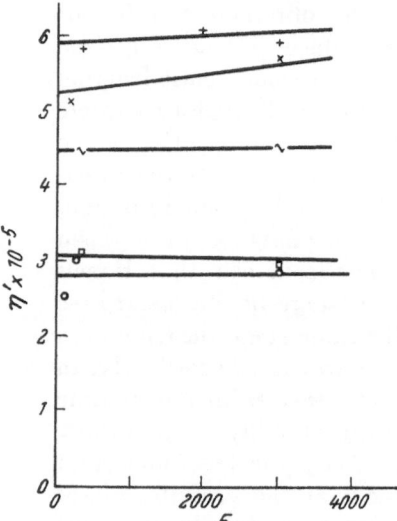

Fig. 3. The viscosity of the protoplasm of the eggs of the sea urchin *Arbacia punctulata*, determined at various shearing forces. These are given by the abscissae in terms of gravities. The ordinates give the viscosity in arbitrary units. The measurements were made at 3—7° C. (Howard.)

A much more complete study of the effect on the protoplasmic viscosity of varying the shearing force was made by Howard (1932). She used eggs of the sea urchin *Arbacia punctulata* and varied the shearing force from gravity to 3,000 times gravity. To obtain the higher forces she used an electric centrifuge. In some of her experiments she corrected for the effect of Brownian movement; for at low shearing forces, the granules tend to return to their original positions. In other experiments she made measurements at low temperatures in order to minimize the effect of Brownian movement. In both sets of experiments the viscosity remained constant in spite of a very wide variation in shearing force. Fig. 3 is a record of the experiments done at low temperatures. Although for individual lots of eggs the viscosity varied markedly, perhaps because of variations in the temperatures at which the measurements were made; for any given lot of eggs the viscosity remained the same over the complete range of shearing forces.

It should be emphasized and emphasized strongly that the viscosity of the protoplasm of marine eggs is not always as low as it is in the cases cited. One reason that protoplasmic viscosity studies are as interesting as they are is the fact that the viscosity of the protoplasm of living cells can change as markedly and as rapidly as it does. In the course of cell division, as will be shown later, the viscosity of the protoplasm rises very rapidly during the prophase (to drop again during the metaphase). As early as 1921, HEILBRUNN noted that the protoplasm of the immature egg of the worm *Nereis* was very stiff, so stiff indeed that with reasonably high centrifugal forces there was no movement of granules or inclusions through the protoplasm.

In his study of the egg of the worm *Sabellaria alveolata*, HARRIS (1935) found the viscosity of the protoplasm of the immature egg (with germinal vesicle intact) to be about 50 times that of the protoplasm in the egg at the time of the metaphase of the first maturation division. And GOLDFORB (1935) states that the viscosity of the protoplasm of *Arbacia* eggs in the germinal vesicle stage is more than 11 times that of the protoplasm of mature eggs—how much more he was unable to determine. The egg of the worm *Chaetopterus* in the germinal

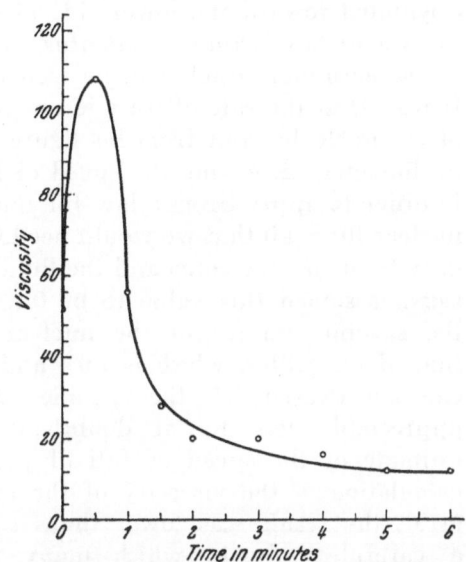

Fig. 4. The viscosity of the protoplasm of the *Chaetopterus* egg at various intervals of time after the egg has entered sea water. Viscosity in arbitrary units.

vesicle stage has protoplasm only 4 or 5 times as viscous as it is after the germinal vesicle has broken down (HEILBRUNN and WILSON 1955 b). This is the condition when the egg has just emerged from the ovary and entered the sea water. But within the space of half a minute, the viscosity goes up markedly and becomes about twice as high. This is shown in Fig. 4 which gives a picture of the viscosity changes which occur in the time immediately following the entrance of the egg into sea water.

Unpublished observations of HEILBRUNN and WILSON show that in the egg of the clam *Spisula solidissima* the viscosity of the protoplasm in the germinal vesicle stage is only a few times as high as it is after the germinal vesicle has broken down. It might be thought that, in view of the relatively high viscosity of the protoplasm in these eggs with germinal vesicle intact, that the protoplasm would show the behavior of a non-Newtonian fluid. However, although our data is meager, what information we have tends to show that this in not true. For when the centrifugal force was varied from 2,250 gravities to 9,000 gravities, as nearly as we could determine, values for the viscosity did not change.

In the case of marine eggs, it is possible not only to obtain values for the viscosity of the cytoplasm; but in some eggs at least it is also possible to determine the viscosity of the fluid within the nucleus, that is to say the nuclear sap. These are the only measurements that have ever been made of the viscosity of the fluid within the nucleus, although in the past various observers have described rapid Brownian movement within the nuclei of plant and animal cells.

As long ago as 1895, Herrick in studying sections of the ovary of the lobster noted that in every case the nucleolus of an egg was always orientated toward the lower side of the nucleus. In the immature egg of the sea urchin *Echinus esculentus,* according to Gray (1927), the nucleolus of the germinal vesicle can be seen to fall under the influence of gravity. It travels at the rate of 0.4 microns per second. Gray did not state the size of the nucleolus, but from his figure it can be seen to be about 16 microns in diameter. Knowing the speed of fall of the nucleolus and its diameter, in order to apply Stokes' law for the determination of the viscosity of the nuclear fluid, all that we would need to know is the difference in the specific gravity of the nucleolus and the fluid through which it falls. In 1928, Heilbrunn assumed this value to be 0.1. This is of course only a guess, but the specific gravity of the nuclear fluid must be a little higher than that of sea water, which is 1.03, and the specific gravity of the nucleolus can not exceed 1.3, the specific gravity of proteins, and is probably appreciably less for it doubtless contains some water. Using Gray's estimate of the speed of fall of the nucleolus, Heilbrunn made a hasty calculation of the viscosity of the nuclear fluid. Due to an arithmetical error, the result was lower than it should have been. Harris (1939) in a careful study in which many thousand measurements were made with a motion picture camera to determine the speed of fall, obtained a value of 9.7 centipoises for the nuclear fluid of *Echinus esculentus.* He used Heilbrunn's estimate of the difference in specific gravity between the nucleolus and the fluid it falls through, and he applied Ladenburg's correction to Stokes' law. Also Harris determined the absolute viscosity of the nuclear fluid by Pekarek's Brownian movement method and found excellent agreement with the value obtained by studying the speed of fall. The value obtained with the Brownian movement method was 9.4 centipoises. As Harris points out, the closeness of the agreement must be fortuitous. One advantage of the Brownian movement method is that it does not require any guesses as to specific gravity.

Only in some eggs is it possible to observe a fall of the nucleolus through the nuclear fluid. Harris tried 13 types of eggs and found he could see a movement of the nucleolus in only 5, all of them representing species of echinoderms. At Woods Hole, the egg of the common starfish can be used for viscosity studies of the nuclear fluid. From data on the speed of fall of the nucleolus, Harding (1949) calculated the viscosity of the nuclear fluid of the immature egg of *Asterias vulgaris* to be 7 centipoises.

Both Harris and Harding found the nuclear fluid to be thixotropic, for the first fall of the nucleolus through the fluid was more rapid than subsequent falls. In a series of 7 experiments, Harris found the average speed of fall in the first test to be about 77 % of that of subsequent falls. In HARDING's experiments, the difference was not quite so great, and HARDING found also that at temperatures above 25⁰ C. the thixotropy disappeared.

A General Statement

It is questionable if any broad general statement can be made about the viscosity of protoplasm. There are so many different kinds of protoplasm in such a vast number of organisms. And in spite of the fact that all these different protoplasms are alike in that they are able to perform essentially the same vital functions; and in spite of the fact also that the chemical composition of various types of living material and the chemical reactions that go on in all different kinds of cells are for the most part very much alike, it would be strange indeed if the viscosity did not vary within very wide limits. For as has already been mentioned, and as we shall see more fully later, the viscosity of any given kind of protoplasm is far from constant. Moreover, the values we now have for protoplasmic viscosity are for only a tiny sample of the myriads of different kinds of living material.

And yet, within certain limits, it is possible to make a few generalizations. In the first place, there are cells in which the viscosity of the protoplasm is only a few centipoises and in which the protoplasm behaves as though it were a true Newtonian fluid. Also there are cells in which the concentration of suspended granular material is so high that the protoplasm as a whole shows the phenomenon of dilatancy, and therefore does not behave as a true Newtonian fluid. However, in these dilatant types of protoplasm, it can safely be assumed that the fluid between the granules has a very low viscosity, for the very fact of dilatancy indicates that this is true. Moreover, although admittedly the number of cells whose protoplasmic viscosity has been tested by proper objective methods is small, nevertheless the measurements do cover rather a wide range of material. Among the cells tested are those of slime molds, algae, the endodermis of bean plants, amebae, ciliates, lobster nerve fibers, frog muscle fibers, the eggs of sea urchins, and the eggs of worms and of clams.

Perhaps it is true that the tests of viscosity have always been made on those types of cells which because of their fluidity lend themselves to such tests. Certainly many cells when they are centrifuged show no movement of the granules they contain. This may be due to the high viscosity of their protoplasm, or it may also be due to the fact that such cells may have an unusually high concentration of granular material. In cells with closely packed granules, there is no room for the granules to move. When a cell has a relatively large nucleus, as in the case of egg cells with a germinal vesicle, the protoplasmic viscosity seems to be especially high. Some tissue cells have nuclei which occupy a large proportion of the total volume of the cell, and it would not be surprising if the viscosity of the cytoplasm of these cells was also high.

Of course what we need now is an extension of our knowledge to as many types of cells as possible. It is not too difficult a matter to separate out cells from tissues and organs and then to study the viscosity of the protoplasm within them. Also one can centrifuge whole organs or whole organisms and then by sectioning discover what effect the centrifugal force has had. This is a rather laborious procedure and has perhaps for that reason been avoided by cell physiologists. The knowledge that we now have is obviously fragmentary and it will be only when we have more information that we will be able to make broad generalizations, if indeed such broad generalizations will ever be possible.

The subject of the viscosity of the outer cortical region of cells, the cortex or ectoplasm, will be discussed by Wilson in another section of this treatise. But at this point is should be pointed out the cortex may perhaps become so thick as to constitute a rather sizable portion of the cell.

The Effect of Various Physiological Factors

Of all the physical variables concerning which we have information or for which we have means of measurement, it is the viscosity of the protoplasm more than anything else which shows changes in relation to the life and death of the cell. At any rate this seems to be true in so far as our present knowledge goes. There is but little evidence concerning changes in refractive index, birefringence, density or zeta potential of the cell as vital processes go on in the protoplasm. It is true that the (thermodynamic) potential difference between the inside and outside of a cell may change when a cell or a part of a cell shows increased activity or is injured, and it is also true that there may also be changes in the osmotic properties of the outer membrane; but even these changes, interesting and important as they are, do not show as close a correlation with vital activity as do the changes in protoplasmic viscosity. Such changes in viscosity occur when a cell grows and ages, when it goes through the process of cell division, when it is injured or dies; they seem to occur when a muscle contracts or when a nerve fiber is aroused, or when, for one reason or another the activity of a cell or its ability to respond is inhibited or suppressed.

The relation of a cell to sudden, sharp changes in its environment, that is to say, to stimuli, is a subject of great importance. It is the outer cortex of the cell, which may or may not constitute a part of the osmotic or plasma membrane, that is of course closer to the environment than is the interior. And thus it is not surprising that when living cells are stimulated by sudden sharp changes in the environment, that it is the cortex which is first and perhaps primarily affected. For example, if a cell is subjected to a sharp mechanical impact, the cortex would immediately be involved. The changes that occur in the cortex as a result of stimulation are of such a nature that they have a strong influence on the cytoplasm in the cell interior.

The essential effect of stimulation on the cortex appears to be a liquefaction accompanied by a release of calcium ion from it into the interior of the cell. The evidence for this point of view and the facts concerning it will be discussed in other sections of this treatise. The nature of the cortex and the physical changes that occur in it will be discussed in the section on the cytoplasmic cortex in another part of this volume. And in Volume VIII, there will be a section on stimulation and anesthesia which will attempt in some detail to explain what happens to the protoplasm when the cell responds to stimulation. At the risk of some repetition, it may not be amiss to point out here that the response of a cell is associated with the viscosity changes that occur as a result of the effect of calcium ions on the viscosity of the protoplasm inside the cell. When a cell divides, the viscosity of the protoplasm always increases during the prophase and then decreases again in the metaphase. This aspect of the subject will be treated in detail in the section on the initiation and prevention of cell division in Volume VI of this treatise. (See also Figs. 4 and 13.)

As cells age there are undoubtedly changes that occur in the viscosity of the protoplasmic colloid, but there have been very few studies of these changes, nor are the studies easy to make. What we would like to know is whether in an aging animal the protoplasmic viscosity of cells in the liver or brain or other organ changes, and how. The only information we have concerns plant cells and marine egg cells, and the aging of marine egg cells is very possibly a different phenomenon from the aging of cells within the animal body.

The literature on the aging of plant cells is in a very unsatisfactory state and there is certainly no unanimity of opinion as to what happens to the protoplasmic colloid as the cells age. This is shown very clearly by FISCHER (1950) in his review of the literature. Although some authors have held that the viscosity of the protoplasm increases with age, others have come to quite the opposite conclusion. STRUGGER (1934) on the basis of plasmolysis measurements, concludes that rapidly growing cells have a higher viscosity than older cells which have ceased to grow. RUGE (1940) gets different results with different methods, and there is a tendency to criticize all types of methods. Even the centrifuge method is open to suspicion, for the specific gravity of the particles moved by the centrifuge may change as the cells age. FISCHER's review should be consulted for references to the literature and for further details. In addition, it may be mentioned that in one of the earliest reports on centrifuged cells, MIEHE (1901) notes—in a single sentence—that inclusions are moved more readily in young cells of onion leaves than in older cells. Similar observations were made on leaf petiole cells of bean plants by MAXIMOV and MOZHAEVA (1944), and they conclude that the older cells have more viscous protoplasm. On the basis of a few centrifuge tests, MODER (1932) reached the same conclusion for *Elodea* leaves. VIRGIN (1951) in some careful studies on *Elodea* leaves, in which he used the centrifuge method to measure viscosity, found that the young cells of the leaf had less viscous protoplasm than cells which were somewhat older. But still older cells showed no further change.

In evaluating the work on plant cells, it should be remembered that the methods that have been used for viscosity determination in most types of plant material do not distinguish between changes in the interior protoplasm and those in the cortex or even in the outer membrane of the cell.

Some of the contradictions in the literature may be due to the fact that different observers made measurements of the viscosity of different parts of the cell.

This is clearly indicated by the recent work of Günther. In her paper (Günther 1957), she not only does experiments of her own on *Elodea* cells, but reviews the older literature. Her conclusion is that with increasing age the interior protoplasm becomes increasingly viscous, but that the cortex (or ectoplasm as she calls it) becomes less viscous. If Günther's judgment is correct, then the data for plant material is in accord with the data for marine egg cells (see below).

Studies of the changes with age of the protoplasm of marine eggs is also difficult. For as the eggs age, the granules they contain fall under the influence of gravity, and before many hours have passed, the eggs appear as though they had already been centrifuged. This is especially true of eggs of the worm *Chaetopterus,* for in these eggs the granules are rather large and they fall more rapidly than do the granules of sea urchin eggs. Heilbrunn (1956 b) tested the viscosity of the protoplasm of *Chaetopterus* eggs as they aged and he tried to prevent the effect of gravity by slowly rotating the eggs. Although it was easy to show that the cortical protoplasm of the egg cells became more fluid as the cells grew older, and this could be shown on eggs which were not rotated, there was some uncertainty as to the changes that might be occurring in the protoplasm in the interior of the egg. This protoplasm at first seemed to become more fluid, but this effect may perhaps have been due to the mechanical agitation caused by rotating the eggs. After 24 hours, the eggs contained many vacuoles. The presence of these vacuoles interfered with the proper determination of viscosity, for the large and relatively light vacuoles were moved by the centrifuge and entered the hyaline zone. However, the vacuolization reaction indicated that the typical clotting reaction of protoplasm had occurred. No doubt the vacuolization is due to the entrance of calcium, and there are indications that when cells of higher animals become senescent, more calcium enters them (Heilbrunn 1956 c).

Goldforb (1935) studied viscosity changes in the protoplasm of aging eggs of the sea urchin *Arbacia*. He used the centrifuge method and his results show steadily increasing viscosity of the protoplasm for some 35 hours, then a liquefaction just before death. In death, the viscosity increased at least 11 times. Goldforb makes no mention of the vacuolization reaction which undoubtedly occurs in aging *Arbacia* eggs. The presence of a few vacuoles may account for what he believed to be a liquefaction before the final great increase in viscosity at death.

According to Weber (1925 e), the viscosity of the protoplasm in the leaves of the aquatic plant *Elodea canadensis* varies with the season. In May and June the viscosity is twice as great as it is in the winter. Moreover, Weber also noted a diurnal variation of viscosity in the protoplasm of the mesophyll cells of the leaves of various species of the genus *Sempervioum*. In the afternoon or early evening the viscosity was relatively low; in the morning it was more than twice as great. Weber made his determinations by observing the time required for chloroplasts to move through the protoplasm under the influence of centrifugal force, and he points out that his results may perhaps be due to changes in the specific gravity of the chloroplasts as a result of a change in their starch content, rather than to changes in viscosity.

The Action of Temperature

In this section and the sections to follow we shall be concerned with the action of various agents on the viscosity of the protoplasmic colloid. If we are to understand the nature and the behavior of this most interesting of all colloids, we need as much information as possible as to how it is changed or influenced by one sort of treatment or another.

Some agents have difficulty in penetrating cells, but in studying the effect of various temperatures on the protoplasm of small cells or groups of cells, one can be sure that the temperature of the protoplasm is the same or essentially the same as the temperature of the environment. But the action of temperature can be complex. For not only is the interior of the cell affected by changes in temperature, but so too is the cortex. And under certain conditions, especially if the change in temperature is great and sudden, what happens to the cortex may affect the interior. This will be discussed more fully later, but before going further it may be well to point out that the suddenness with which the temperature is changed may to some extent at least determine the effect of temperature on the interior protoplasm. Some of the differences in the results obtained by various workers may be due to this fact, although more probably the differences are due to a diversity in the types of protoplasm that have been studied.

The first measurement of the effect of temperature on protoplasmic viscosity was made by FRIEDL and GISELA WEBER in 1916. They used seedlings of the bean plant *Phaseolus multiflorus* and cut sections through the endodermis so as to expose the starch sheath cells. The sections were then placed at constant temperatures on a microscope stage and after an interval of a half hour or more in order to make sure that the effect of wound injury had passed, the WEBERS observed the speed of fall of the starch grains through the protoplasm at various temperatures. Their results are given in the following table. In this table, each row of figures represents an experiment with a single cell. Comparisons should be made for each cell individually, as the different cells vary.

Clearly, the viscosity, which is proportional to the time required for the fall of the starch grains, decreases steadily as the temperature is raised, and the WEBERS point out that this decrease follows the same course as the decrease in viscosity of protein solutions with rising temperature.

The fact that the protoplasm is still fluid at 60° C. is rather strange, and the WEBERS point out that this is true only for the sections. In the intact plant the protoplasm ceases to be fluid at much lower temperatures. Also it should be noted that these values of the WEBERS are values for the entire protoplasm, rather than just for the granule-free protoplasm.

A similar decline of protoplasmic viscosity with increasing temperature was found also for the protoplasm of the egg of the nematode worm *Ascaris megalocephala* by Fauré-Fremiet (1913) in what was one of the first

Time, in seconds, required for the fall of starch grains through the protoplasm of Phaseolus cells

Temp.C.0°	4°	8°	10°	14°	20°	24°	30°	34°	40°	44°	50°	54°	60°
			8		6								
			14		11.5								
							10		8.5		7.5		
							14		11		8		7
							6		5		4		3.5
			20			16		10		7			
	16	14	12	11	8		4						
			20		12		7	6					
			32		18		14		10		9		
					13		10		7		6		5
21			15		11		8						
			12		9		7		5				
20			15		10		7.5		6				
	9	7	6.5	5									
		11	10		8		7		6		5		
			8.5		7		6		5				4.5
		16		14		12		9.5		8		6.5	

measurements of protoplasmic viscosity with the centrifuge method. With a centrifuge that made 2.500 turns per minute, Fauré-Fremiet determined the length of time required to move the mitochondria of the egg from one end of the cell to the other. (He does not give the radius of turn of the centrifuge.) His results are shown in the following table.

The viscosity of Ascaris egg protoplasm at different temperatures

Temperature	Time (i. e. relative viscosity)
8° C.	225 mim.
18°	45
23°	20
30°	8
35°	4

Below 8⁰ C. the viscosity was infinite and even at 8⁰ C. not all the mitochondria moved in the time given. It should be remembered that the normal temperature for the egg of *Ascaris megalocephala* is above 35⁰ C., for its habitat is the intestine of a horse. Apparently the movement of mitochondria that FAURÉ-FREMIET observed was through a rather heavy concentration of suspended material, so that in this case also we are concerned with the viscosity of what is essentially the entire protoplasm.

It must not be supposed that the viscosity of protoplasm always shows a progressive decrease as the temperature is raised. Thus HEILBRONN (1922) in studying the viscosity of the protoplasm of the slime mold *Reticularia lycoperdon* with his magnetic method was surprised to find that as he lowered the temperature from 17⁰ C. to 12⁰ C., the viscosity instead of increasing, decreased approximately 10 percent. This was true in each of three separate experiments. The effect was not a sudden, transient one, for in one case at least, the slime mold was exposed to the lower temperature for nearly two hours. On the other hand when the temperature was raised above 17⁰ C. the viscosity also decreased. Thus in slime mold protoplasm as the temperature is raised the viscosity goes through a maximum. In this work of HEILBRONN'S also, it is the viscosity of the entire protoplasm that was measured.

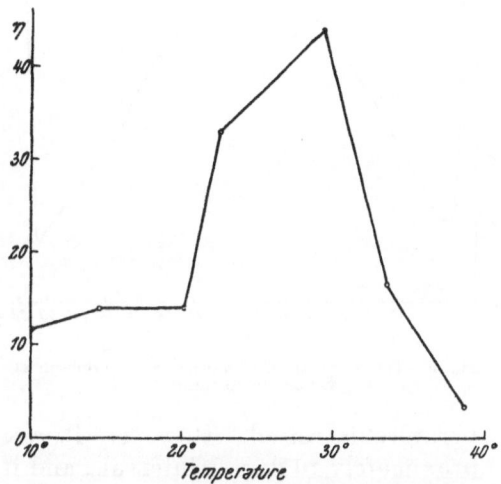

Fig. 5. The mean viscosity of *Spirogyra* protoplasm at different temperatures. The ordinates show viscosity in centipoises.

(After BAAS-BECKING, SANDE BAKHUYZEN, and HOTELLING.)

In their study of the protoplasm of the alga *Spirogyra*, a study in which they made numerous measurements of Brownian movement, BAAS-BECKING, SANDE BAKHUIZEN, and HOTELLING (1928) thought that the viscosity was highest at about 27⁰ C. This is illustrated in Fig. 5 which shows the mean viscosity values obtained at different temperatures. The peak in this curve is at 29.3⁰ C. These viscosity determinations made as a result of Brownian movement measurements are concerned primarily with the viscosity of the hyaline, granule-free protoplasm; and although, as pointed out in an earlier section, there are errors involved in the measurements, it seems probable that in this type of protoplasm, the viscosity goes through a maximum as the temperature is raised.

HEILBRUNN (1924 a) studied the viscosity of the protoplasm of the egg of the clam *Cumingia* at different temperatures from 0 to 32⁰ C. Determinations were made by the centrifuge method. The results are shown in Fig. 6. The *Cumingia* protoplasm shows a maximum viscosity at about 15⁰ C. Above and below this temperature the viscosity is less. However the viscosity becomes markedly higher when the temperature falls to 1⁰ C. or when it is raised to 32⁰ C. In this work of HEILBRUNN'S no mention is made of the length of time the eggs were exposed to one temperature or another. It might be supposed therefore that when eggs were suddenly exposed to either a low a high temperature, the sudden change might have something of a shock effect. In the *Chaetopterus* egg, sudden heat and sudden cold both cause a liquefaction of the cortex and this presumably is accompanied by a release of calcium (WILSON and HEILBRUNN 1952). However such an effect causes not a decrease but rather an increase in protoplasmic viscosity (HEILBRUNN and WILSON 1955 a). Also it should be noted that the viscosity values given by HEILBRUNN in his 1924 a paper represent a series of tests over a period of time which in general was at least 20 minutes. Thus it can scarcely be assumed that the drop in viscosity on either side of the maximum is due to a shock effect.

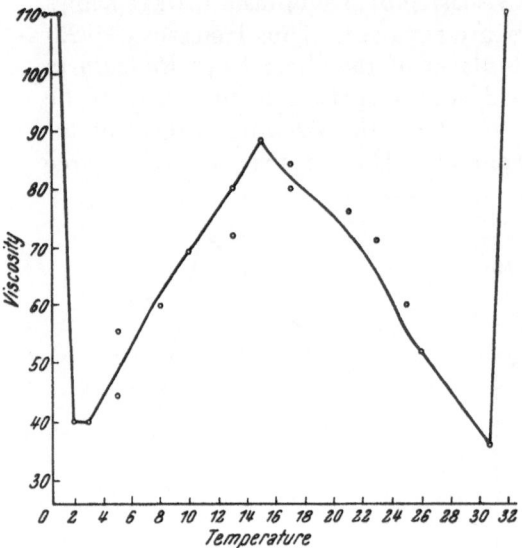

Fig. 6.   The viscosity of *Cumingia* egg protoplasm at various temperatures.

PANTIN (1924) studied the effect of temperature on the viscosity of the protoplasm of the immature eggs of the worm *Nereis versicolor*. His results are shown in Fig. 7. They show clearly a progressive decrease in viscosity as the temperature is raised. However, PANTIN's measurements were made at approximately 10 degree intervals, and it is conceivable (though not probable) that there might be a maximum in the viscosity between 10⁰ and 20⁰ C. and minima in the neighborhood of 5⁰ and 20⁰ C.

A more complete study is that of COSTELLO (1934). Using a force 2,618 times gravity, he centrifuged *Arbacia* eggs at temperatures from 1.5⁰ to 28⁰ C. His measurements required from one to nearly three minutes, and with the centrifugal force he used, the temperature would ordinarily tend to rise rapidly during the course of the centrifuging.

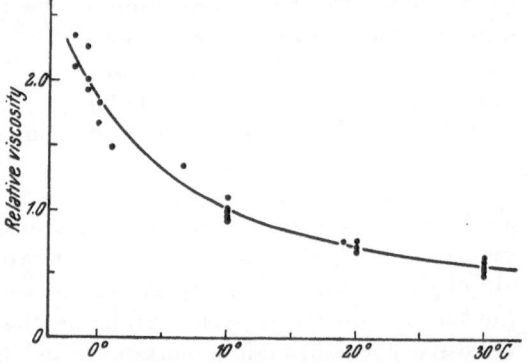

Fig. 7.   The relative viscosity of the protoplasm of *Nereis* eggs at different temperatures.
(PANTIN.)

COSTELLO was careful to control this factor by passing water at any desired temperature through the chamber surrounding the head of the centrifuge. Eggs were exposed for 30 minutes to the experimental temperature before measurements were made. COSTELLO's data are shown in Fig. 8.

In *Amoeba dubia*, HEILBRUNN (1929 b) found that the viscosity of the protoplasm in the interior decreased sharply as the temperature was increased from 3⁰ to 18⁰ C. Then there was a rise in viscosity until a maximum was reached at around 25⁰ C.—above this temperature the viscosity

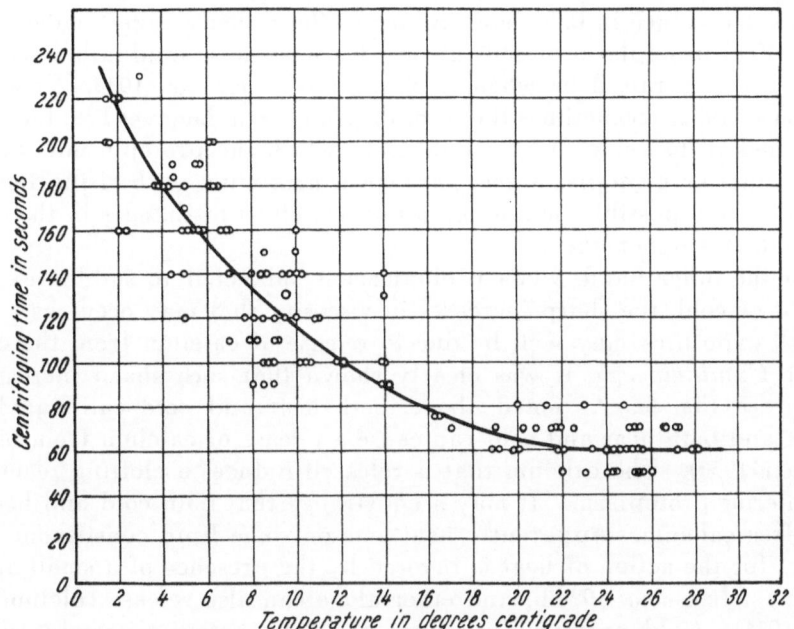

Fig. 8. Viscosity of the protoplasm of eggs of *Arbacia punctulata* at different temperatures. The ordinates show viscosity in terms of the time required to produce a hyaline zone one sixth the diameter of the egg. Each point represents a different lot of eggs.
(COSTELLO.)

decreased again. This is shown in Fig. 9. MURPHY (1940) repeated HEILBRUNN's study of the effect of temperature on the viscosity of the protoplasm of *Amoeba dubia*, but he made no measurement above a temperature of 20.8⁰ C. His results show a sharper drop of viscosity between 3⁰ and 18⁰ C. due to the fact that he obtained somewhat higher values at the lower temperatures. These values for the viscosity at the low temperatures are more accurate than those of HEILBRUNN, for MURPHY was careful to do his centrifuging at the same low temperatures to which the amebae were exposed.

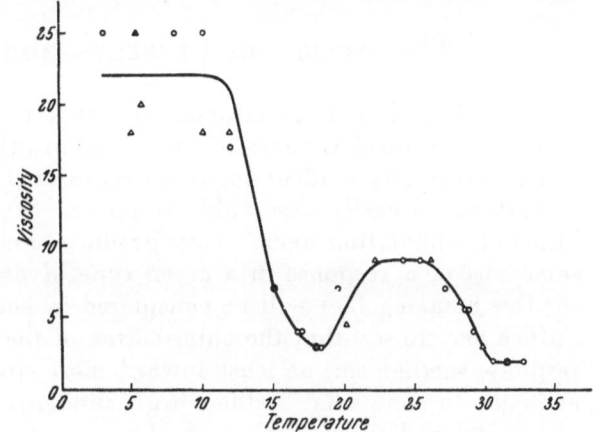

Fig. 9. Viscosity of the protoplasm of *Amoeba dubia* at different temperatures.

This survey of the effect of temperature on protoplasmic viscosity seems to indicate rather clearly that there are two types of behavior. The protoplasm of some cells shows a progressive decrease in viscosity as the temperature is raised—in other cells the viscosity goes through a maximum with rising temperature. At present it is not possible to state the reason for this divergence in behavior. As far as the rather meager evidence goes, the cortical protoplasm of cells always becomes more fluid either when the temperature is raised or when it is lowered (THORNTON 1935, WILSON and HEILBRUNN 1952). Sometimes the interior protoplasm behaves like the cortex, sometimes it does not. In those cases in which the protoplasmic viscosity of the interior protoplasm goes through a maximum with rising temperature, this may possibly be due to an indirect effect of changes in the cortex, although this is not likely.

On the other hand, when protoplasm is subjected to sudden extremes of heat or cold, the sharp increases in viscosity that may occur as a result of such exposures may well be due to release of calcium from the cortex. In the *Cumingia* egg, it was clearly shown that such sharp increases do occur; see Fig. 6. As noted above, both heat and cold can liquefy the cortex and both heat and cold can cause a release of calcium from binding (WEIMAR 1953). The calcium that is released induces a clotting reaction in the interior protoplasm. It may seem strange that both cold and heat can both free calcium. Apparently heat acts on some lipid constituent of the cortex, for the action of heat is favored by the presence of a small amount of ether (HEILBRUNN 1924 b) and ether alone can also release calcium (BERWICK 1931). Cold presumably acts on the bond between protein and calcium, for the cation binding power of a protein decreases with a decrease in temperature (AUSTIN, SUNDERMAN, and CAMACK 1927). This point of view thus presents a plausible interpretation of why it is that cold and heat sometimes produce the same sorts of effects on biological systems—both can act as stimulants, both can act as anesthetics.

The Action of Pressure and of Electricity

One of the essential characteristics of living protoplasm is its ability to respond to sudden changes of environment—that is to say, to stimulation. This is especially evident for those types of protoplasm which are capable of making an easily observable response. Strangely enough, very diverse kinds of stimulating agents may produce—and typically do produce—the same sort of a response in a given type of protoplasm. The basic reason for this amazing fact will be considered in another section of this treatise. Suffice here to say that the outer cortex of the cell apparently initiates the response mechanism, at least toward most stimuli. There is considerable evidence to show that sudden heat, sudden cold, sharp uneven pressure, ultraviolet radiation, electric shocks, as well as certain types of chemical treatment, all cause a release of calcium from the cortex, and that this calcium on entering the cell interior causes a sharp increase in protoplasmic viscosity.

If a cell is subjected to uneven pressure or to an electric shock, the effect produced on the cell as a whole is so much a result of the release of calcium from the cortex that it is at the present time impossible to state what the effect on the interior would have been if the cortex had not been involved.

The entering calcium causes first a lowering of viscosity, for, as will be noted later, the entrance of small amounts of calcium into the interior of a cell causes a drop in the viscosity of the protoplasm there. Then as more calcium enters, the viscosity of the protoplasm increases sharply, for calcium has a clotting effect on some constituents of the protoplasm. The course of these changes is clearly shown in Fig. 10, which is taken from the

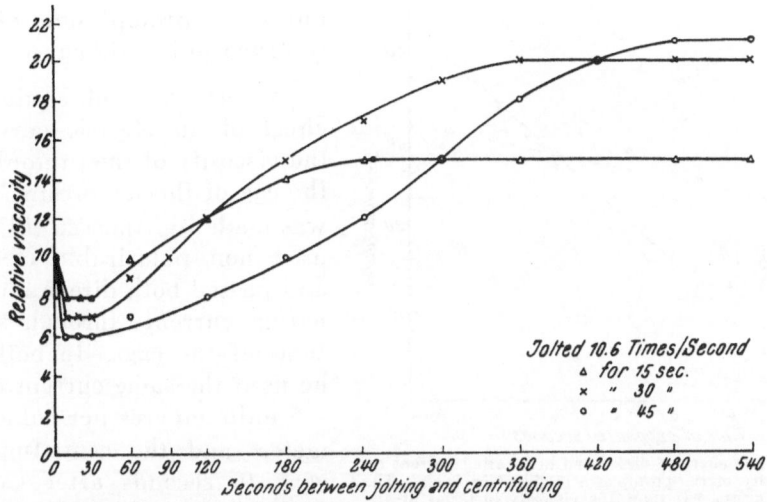

Fig. 10. The relative viscosity of the interior protoplasm of *Amoeba dubia* following exposure to shaking for various lengths of time. Measurements were made with the centrifuge method.
(ANGERER.)

work of ANGERER (1936). He stimulated amebae by bouncing them around in a shaking machine. Then he studied the viscosity both of the outer cortical protoplasm and the protoplasm in the interior. The viscosity of the cortical protoplasm dropped sharply as a result of the mechanical stimulation, an indication that calcium was released from it. On the other hand, the viscosity of the protoplasm in the interior of the ameba first dropped appreciably and then increased.

The fact that a very small electric current can produce a marked effect on protoplasmic viscosity is a striking example of the unique behavior of the protoplasmic colloid, for inanimate colloids do not behave in this way. That electric currents do act on protoplasmic viscosity has been known for nearly a hundred years. In 1862, BRÜCKE showed that electric currents caused a cessation of Brownian movement in the leukocytes found in human saliva. And two years later KÜHNE (1864) noted that the passage of an electric current from an induction coil stopped Brownian movement in the stamen hair cells of *Tradescantia*. Similar observations were made by CHIFFLOT and GAUTIER (1905) for the protoplasm of the alga *Cosmarium* and by BAYLISS (1920) for the protoplasm in the pseudopodia of an ameba.

BERSA and WEBER (1922) made a careful study of the effect of an electric current on the protoplasm of the bean plant, *Phaseolus multiflorus*. They used starch sheath cells from the epicotyl of the bean plants and measured viscosity by centrifuging the starch grains through the protoplasm and noting the time required for the movement. Relatively strong currents —5 to 10 milliamperes—produced an effect in 15 to 30 seconds; whereas weaker currents—0.15 to 5 milliamperes—required a longer time (1 to 4 minutes). Both weak and strong currents produced an increase in the viscosity. This was approximately a threefold increase. The effect was reversible and after 20 to 40 minutes, the protoplasmic viscosity returned to its original low value.

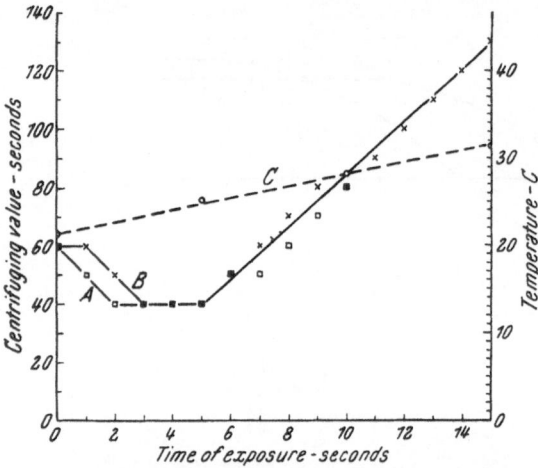

Fig. 11. The effect of direct and alternating current on the viscosity of the protoplasm of the *Arbacia* egg. The ordinates at the left show the viscosity in terms of the number of seconds required to shift the granules of the cytoplasm a given distance. Curve *A* gives the results for alternating current; curve *B* for direct current. In both cases current density was 0.005 amperes/mm². Curve *C* shows the change in temperature of the trough through which the current passed.
(ANGERER.)

A very thorough study of the effect of an electric current on the viscosity of the protoplasm of the egg of the sea urchin *Arbacia* was made by ANGERER (1939). He used non polarizable electrodes and passed both direct and alternating currents through suspensions of the eggs. In both cases he used the same current density —5 milliamperes per cubic millimeter—and the centrifuged the eggs 10 seconds after exposure with a force approximately 2,500 times gravity. The time of exposure to the currents varied from 1 to 15 seconds. ANGERER's results are shown in Fig. 11.

Both in the case of the direct and the alternating current, the first effect of the electricity is to cause a transient decrease in viscosity; the viscosity drops to two-thirds of its original value. This is followed by a progressive increase in viscosity. Thus the changes are similar to those shown to occur in ameba protoplasm following mechanical stimulation. Moreover, as in the case of mechanical stimulation, electric currents cause a drop in the viscosity of the cortical protoplasm—at any rate this is true for the ameba (ANGERER and WILBUR 1943).

Radiation Effects

Visible Light

For the most part, the effects of visible light on protoplasmic viscosity have been studied by botanists. The reason for this is quite obvious; plant protoplasm is apparently much more sensitive to visible light than is the protoplasm of animal cells. In interpreting the results that have been reported, caution is necessary. If an observer finds that the chloroplasts of a plant cell move more readily or less readily following illumination, this does not necessarily imply a change in viscosity. As is well known, chloroplasts readily change their position within the protoplasm. If under one condition they are close to or imbedded in the cortical protoplasm and under another condition they move away from the cortex, such movement might have a profound effect on the ease with which chloroplasts could be displaced by centrifugal force.

Moreover, as pointed out in the section on the measurement of protoplasmic viscosity, botanists have almost never distinguished between the viscosity of the interior fluid protoplasm and the viscosity of the more or less rigid cortex. It is true that HEILBRONN (1922) in studying the large masses of protoplasm in slime molds was able to make such a distinction, but ordinarily in plant cells the protoplasmic layer surrounding the central vacuole is so narrow that no distinction between cortex and interior protoplasm is possible; at any rate such a distinction has not commonly been made. In the earlier section on methods, it was noted that the displacement of chloroplasts by the centrifuge in cells like those of the alga *Spirogyra* is to be regarded not as a measure of the viscosity of the fluid protoplasm of the cell, but to some extent at least as a measure of the force necessary to break loose these structures from their moorings in the outer cortical region of the protoplasm. As has repeatedly been demonstrated for animal cells (HEILBRUNN 1952, 1956 a), the properties of the stiff cortex and its viscosity changes are often very different from those of the interior protoplasm.

In 1925, WEBER (1925 e) recognized the fact that light might have an effect on the protoplasmic viscosity of leaf cells, and in 1927, using his plasmolysis method, he found marked differences in the behavior of stomatal cells in leaves of the broad bean (*Vicia faba*) when these were in the light and when they were in the dark. The conclusion he reached was that the light had a marked effect in increasing the viscosity of the outer layers of the protoplasm, that is to say the cortex. Also he noted that protoplasmic streaming was much more vigorous in the dark than in the light, an indication that the interior protoplasm also suffered an increase in viscosity on exposure to light. Later (1929 b) in experiments on leaves of the plant *Ranunculus ficaria* L., he again found from plasmolysis studies that the viscosity, presumably of the cortex, was higher in the light than in the dark.

More recently the effect of light on protoplasmic viscosity has been thoroughly investigated by STÅLFELT and VIRGIN, and both of these authors have reviewed the literature (VIRGIN 1953, STÅLFELT 1956). STÅLFELT (1946) tried to find out what the effect of light was on the protoplasm in the leaves of *Elodea densa*. He centrifuged the leaves and determined the length of time it took for the chloroplasts to move through the protoplasm. In his first experiments, STÅLFELT ran into serious difficulties, for when the leaves were subjected to centrifugal force, the time required for the chloroplasts to move varied widely. But then he discovered that if the leaves were kept in the dark for three days and were then cut off in a weak red light and centrifuged, the time required for the chloroplasts to move was the same for all the leaves and remained constant for several hours. Stålfelt believes that the time required for the movement of these chloroplasts is a true measure of protoplasmic viscosity. After the leaves had been kept in the dark for three days and were then exposed to weak light (5 to 4,000 meter candles), the viscosity rises at first and then falls again. Equilibrium is not reached and periodical changes in viscosity occur. In stronger light (8,000 to 16,000 meter candles), the viscosity at first decreases and then increases, and then various fluctuations follow. With still greater light intensity (32,000 to 48,000 meter candles), the viscosity decreases at first but then remains constant or nearly so. Stålfelt believes that "if the electrical charge on the particles is influenced by the light this will cause a change in the viscosity of protoplasm." And he goes on to say "the way in which the light influences the electrical charge of the particles is unknown." However, Heilbrunn and Daugherty (1939) showed clearly that when *Elodea* cells had been kept in the dark, in the absence of photosynthesis the electric charge on the chloroplasts (and presumably on other colloidal particles of the protoplasm) was in all cases positive. In the light, on the other hand, the charge could be either positive or negative. Presumably the sign of the charge and its magnitude is affected by the concentration of carbon dioxide in the protoplasm. This work of Heilbrunn and Daugherty is mentioned by Virgin and he believes that the "charges may be invisaged as affecting the displacement of the chloroplasts within the cells, but the mechanism is uncertain."

There are various ways in which an electric charge on particles could affect the viscosity as determined by the centrifuge method. For colloidal particles of ultramicroscopic size, that is to say for particles so small that they can not be seen with a light microscope, an electric charge would tend to hinder their movement under the influence of centrifugal force, for the charge would tend to cause a decrease in viscosity (von Smoluchowski 1916). This is the well known electroviscous effect (see Overbeek and Bungenberg de Jong 1949). But in a gross suspension of high concentration, the greater the charge the more fluid the suspension as a whole would tend to become, for the charge would tend to prevent the suspended particles from clumping together. Were the charge neutralized, clumping or clotting would occur and this would tend to retard the passage of a larger inclusion like a chloroplast and it might also retard the movement of individual granules. At any rate, increase in charge on the granules of a protoplasmic suspension seems to decrease the viscosity (compare Heilbrunn 1923, 1928). In view of these facts it might be possible, to same extent at least, to interpret the effects of light on the viscosity of the protoplasm of *Elodea* leaves. For as the charge on the chloroplasts was neutralized by the light and then reversed, various types of viscosity changes might occur. Probably this is not the only factor involved, but it may well be an important one.

Presumably because of the insensitivity of the protoplasm of animal cells to visible light, there has been very little study of the effects of such light on animal cells. Recently, however, there has been considerable interest in the phenomenon of photoreactivation, for this occurs not only in bacteria and fungi, but also in protozoa and in sea urchin eggs. In photoreactivation, the changes produced by ultraviolet radiation of wave lengths around 2,500 A are counteracted at least to some extent by exposure to radiation of longer wave lengths in the range between 3,300 and 4,800 A. The subject has been reviewed by DULBECCO (1955). He believes that the primary action is on the nucleus and involves changes in nucleic acid. However, as will be shown later, beyond any question ultraviolet radiation produces changes in the cytoplasm as well as those it might cause in the nucleus. And in so far as the cortical protoplasm of ameba is concerned, the changes produced by ultraviolet radiation are exactly the opposite of those produced by visible light. This follows from the work of HEIL-BRUNN and DAUGHERTY (1933) and of ALSUP (1942). In the ameba, as will be discussed more fully later, ultraviolet radiation causes a marked decrease in the viscosity of the cortical protoplasm. On the other hand, ALSUP found that strong visible light had quite the opposite effect. He used very intense illumination. With a 1,000 watt incandescent lamp, he exposed amebae to approximately 300,000 meter candles. In this work he was careful to exclude heat by using a water filter as well as a fan. Following exposure to the light, in 17 experiments there was a consistent and a pronounced increase in the viscosity of the cortical protoplasm, an increase which averaged 34%. This increase was dependent on the presence of calcium, for if the amebae were previously exposed to dilute solutions of ammonium oxalate, the light did not have any effect on the viscosity. Here then is a clear case in which visible light acts on the cytoplasm in a manner directly opposite to that in which ultraviolet light works. And this phenomenon should be taken into account by those who speculate about the reasons for photoreactivation.

ALSUP also studied the effect of visible light on the viscosity of the protoplasm in the interior of *Amoeba dubia*. On the average, the viscosity of this protoplasm doubled when the amebae were exposed to approximately 300,000 meter candles of illumination. This viscosity increase also seemed to be dependent on the presence of calcium, for when amebae were previously exposed to dilute solutions of ammonium oxalate, the light was without effect.

The work of ALSUP on ameba protoplasm may well be correlated with the early work of WEBER on various types of plant cells. For both ALSUP and WEBER believe that visible light causes a marked increase in the viscosity both of the cortical protoplasm and of the protoplasm in the interior. Any attempt to interpret this effect of visible light should take into account the fact that visible light also causes a very definite increase in the viscosity of the cell sap of plant cells (PEKAREK 1933).

Ultraviolet Radiation

Much has been written about the effect of ultraviolet radiation on living systems, and indeed all types of protoplasm appear to be sensitive to ultraviolet rays. Recent reviews include an entire volume—volume 2— of HOLlaender's Radiation Biology, published in 1955, and a discussion by Errera (1957) which forms a part of this treatise.

The rather common opinion—and it has been held for some time—is that the primary and the essential effect of ultraviolet rays on a living cell is an effect on the nucleus. There is much evidence in support of this view. For one thing, nucleoproteins absorb certain wave lengths in the ultraviolet strongly, and before it was realized that the cytoplasm also contains nucleoproteins this was sometimes urged as a reason for emphasizing the importance of the nucleus in the interpretation of ultraviolet effects. There seems to be no very clear idea as to how the nucleus is affected by the radiation, or how once it is affected it can influence such cytoplasmic phenomena as the division of a cell. For the time relations of cell division are determined primarily by the action of the mitotic spindle and its asters; and so if ultraviolet radiation retards the course of cell division, this can not be due to an action on the nucleus alone.

Giese (1946) investigated the action of ultraviolet rays of different wave lengths on sea urchin eggs and he found that a wave length of 2,800 A was most effective in retarding cell division. This would indicate that simple proteins rather than nucleoproteins were involved. But when spermatozoa were irradiated and then used to fertilize non-irradiated eggs, the action spectrum indicated that nucleoproteins were primarily affected. Thus there may well be a double effect. We must always remember that nucleus and cytoplasm undoubtedly interreact, so that whatever happens to the nucleus may affect the cytoplasm and whatever changes occur in the cytoplasm may affect the nucleus.

In this connection it is interesting to consider an experiment reported by Harding and Thomas (1950). They exposed centrifuged eggs of *Arbacia punctulata* to ultraviolet radiation of wave length 2,537 A. The eggs were oriented in a clever way so that simultaneously one lot of eggs was irradiated with rays that entered the cell close to the nucleus, whereas another lot of eggs received the same dose from rays that entered the cell at a distance farthest from the nucleus. Both sets of eggs had their cleavage delayed, but the delay was somewhat greater when the rays entered the cell close to the nucleus. This suggests the possibility that there may be an effect on the outer cortical protoplasm, but that the result of this action is greater the nearer this layer of protoplasm is to the nucleus. As will be shown later, we have good reason to believe that the cortical protoplasm of a cell is affected by radiation, and that following exposure to the rays, the cortex liquefies and releases calcium. We know also that this released calcium can activate a proteolytic enzyme which apparently is responsible for the destruction of the nuclear membrane, an essential step in the division of the cell (Goldstein 1953). Also, according to Goldstein, the proteolytic enzyme by its digestive action releases an inhibitor to its own activity. This train of ideas might well explain why it is that when ultraviolet rays strike a cell near its nucleus they might produce a more powerful action than when they first strike the cell at some part remote from the nucleus.

Much of the argument concerning the all-importance of the nucleus in irradiation effects neglects the possible role of interactions between nucleus and cytoplasm. As a matter of fact such interactions were clearly recognized by SCHLEIP (1923), whose experiments and conclusions have never been given the attention they deserve. SCHLEIP centrifuged *Ascaris* eggs and then irradiated localized regions of the centrifuged cells with ultraviolet radiation of 2,800 Å. He was able to irradiate granule-free protoplasm. light or heavy granular material, and also that part of the cell which contained the nuclei (although of course in irradiating the nuclei, the rays had to pass through the cytoplasm on their way into and out of the nuclei). SCHLEIP concluded that injury to the cytoplasm could affect the nucleus. When he irradiated the nucleus (plus cytoplasm in the path of the rays), the dosage necessary to produce an effect was much lower than when he irradiated parts of the cytoplasm, but the effect when he irradiated cytoplasm alone was by no means negligible.

In a recent experiment on immature eggs of the starfish *Asterias vulgaris,* HARDING (1955) has presented additional evidence to show that the effect of ultraviolet radiation on the nucleus is in part at least due to changes in the cytoplasm. For when he irradiated eggs so that no part of the nucleus was in the path of the rays, the nucleolus broke down. Also from his studies of such partial exposures and from a mathematical calculation of the results obtained, he found that the effect on the nucleus was dependent rather on the extent of the surface area hit than on the volume of protoplasm in the path of the rays. This led to the conclusion that the cortex of the egg plays an unusual role. A similar conclusion might have been reached from the earlier work of HEILBRUNN and WILBUR (1937), work which indicates that the breakdown of the germinal vesicle in the egg of *Nereis* following exposure to ultraviolet radiation is dependent on the presence of calcium in the outer region of the egg.

GIBBS (1926) studied the effect of ultraviolet radiation on cells of the alga *Spirogyra*. He exposed the cells to radiation from a mercury vapor lamp which emitted rays of wave length 2,378 to 3,126 Å and then centrifuged the cells for varying lengths of time. Relatively short exposure to the radiation caused the spiral chloroplasts to be displaced more readily than they were in the control untreated cells. Longer exposures had the opposite effect, and GIBBS concluded that the shorter exposures caused liquefaction, whereas the longer exposures caused coagulation. How much of this was due to a change in the cortex, the interior protoplasm, or both, is uncertain.

The experiments of VOERKEL (1933) can perhaps be interpreted with a little more assurance. He worked with cells of the moss *Funaria hygrometrica* and in these cells the relatively small chloroplasts are presumably not so well anchored as are the large chloroplasts in *Spirogyra* and are perhaps free in the protoplasm. Relatively low doses of ultraviolet radiation caused a marked decrease in the viscosity of the protoplasm as determined by the centrifuge method. Stronger doses had a coagulative effect.

Biebl (1942, 1947) exposed epidermis cells from the scales of onions to the radiation from a mercury vapor lamp and noted what he believed to be effects on the protoplasmic viscosity. Following exposure, he centrifuged the cells in a centrifuge that gave 2,800 revolutions per minute for 1 and 2 hours. After this long treatment, the irradiated cells had their protoplasm thrown to one side of the cell, the controls did not. This can perhaps be taken as evidence for a decrease in viscosity and so Biebl considers it. More certain evidence is the fact, also noted by Biebl, that after radiation there is active Brownian movement of small granules in the protoplasm.

Schleip (1923) and Ruppert (1924) in their studies of the effect of ultraviolet radiation on the development of *Ascaris* eggs presented clear evidence of a sharp increase in protoplasmic viscosity as a result of the exposure of the eggs to radiation of 2,800 Å wave length. For when, following such exposure, the eggs were centrifuged, the return movement of the granules in the treated eggs was much slower than in the control eggs and commonly this return movement did not occur at all in the treated eggs.

Heilbrunn and Young (1930) studied the effect of ultraviolet rays on eggs of the sea urchin *Arbacia punctulata*. These eggs are stimulated by the radiation; they undergo the membrane elevation that normally is a sequence of fertilization and many of the eggs start cleavage and begin to develop. Excess radiation results in a reaction called cytolysis by students of marine eggs; in this reaction the protoplasm becomes filled with vacuoles. Five or six minutes after a one minute exposure to the radiation from a mercury vapor lamp, the viscosity of the protoplasm decreased by about 30% (in terms of absolute viscosity, from 3 to 2 centipoises). Then some 15 minutes after the exposure the viscosity rose until it was several times as great as it was in the eggs before exposure. These changes seemed to proceed from the outer peripheral part of the egg toward the interior. (But Heilbrunn and Young had no information about the cortex of the egg, for the existence of a thin cortex in the unfertilized sea urchin egg was not understood until the work of Moser in 1939.) The changes that occur in the protoplasm of the sea urchin egg as a result of the exposure to ultraviolet radiation depend on the presence of calcium. If eggs are first treated with oxalated sea water and then irradiated, no viscosity changes occur, and there is no response to the irradiation, no membrane elevation and no cleavage. And yet the eggs are not injured by the oxalate, for on return to sea water they show the normal response to irradiation.

A more complete study of the effect of ultraviolet radiation on protoplasmic viscosity was made by Heilbrunn and Daugherty in 1933. They exposed ameba to the radiations from a mercury vapor lamp. As could be determined by tests of viscosity with the centrifuge method, these radiations cause a sharp decrease in the viscosity of the cortical protoplasm of *Amoeba proteus,* and this Heilbrunn and Daugherty believed to be correlated with the fact that ultraviolet radiation causes a release of calcium from the cortex. The effect of the radiation on the fluid protoplasm in the interior of the ameba was studied by centrifuge tests made on *Amoeba dubia*, for in this species the cortex is very thin and does not obscure observation of the interior. When the amebae were exposed for 2 minutes, a total of 16 experiments gave consistent results; these experiments showed an average viscosity increase of about 400%. However, the viscosity increase is preceded by a transient drop in viscosity. This is illustrated in Fig. 12 which shows what happens when amebae were exposed for 5, 10, 15 and 20 seconds. The lower doses caused only a temporary viscosity decrease, the higher doses a preliminary decrease followed by a sharp increase in viscosity. The effect is what one might expect if calcium were released into the interior from the cortex. For as will be

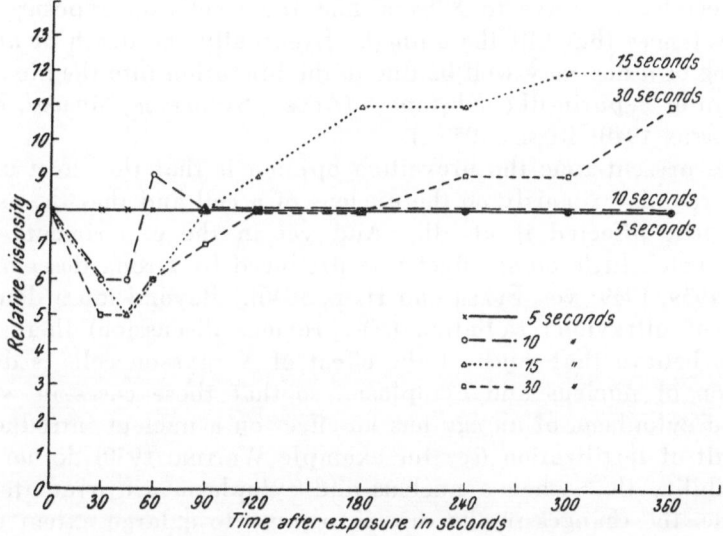

Fig. 12. Changes in the viscosity of the interior protoplasm of ameba following brief exposures to ultra-violet radiation. The various curves represent different lengths of exposure.

shown in a later section, calcium in low concentration causes a decrease in protoplasmic viscosity, whereas in higher doses it causes clotting or gelation.

These results on various types of protoplasm are all reasonably consistent. Relatively low intensities of ultraviolet radiation cause a decrease in protoplasmic viscosity, larger doses have the opposite effect. With the larger doses there may be a transient decrease in viscosity before the viscosity increase begins.

Finally, perhaps mention should be made of an amazing paper by SHIRLEY and FINLEY, published in 1949. These authors measured the viscosity of the protoplasm of the ciliate *Spirostomum ambiguum* by noting the length of time it took for the organisms to disintegrate when they were dropped on a slide which was covered with crystals of NaOH! It is hard to conceive of a less scientific method of attempting to study viscosity, and yet in his review on the effects of radiation on protozoa and invertebrate eggs other than those of insects, a review published in Hollaender's authoritative treatise on Radiation Biology, this is the only paper on the effect of ultraviolet radiation on viscosity that is cited by the reviewer (Kimball 1955).

X Rays and β Rays

Cells and organism differ enormously in their sensitivity to X rays. Whereas it takes only a few hundred r units to kill a man or a horse, a paramecium can be exposed to hundreds of thousands of r units and still live. The reason for this remarkable diversity in behavior is not at all understood. All that one can say is that in higher animals there are some cells especially sensitive to X rays, and these cells on exposure give off toxic substances that kill the animal. Eventually the death of an animal like a frog or a dog may well be due to the liberation into the blood stream of heparin or heparin-like substances (Allen, Sanderson, Milham, Kirschon, and Jacobson 1948; Rieser 1955 a).

At the present time the prevailing opinion is that the effect of X rays and of β rays is primarily on the nucleus of a cell and that the cytoplasm is very little affected if at all. And yet in the experiments done on isolated nuclei, little or no effect was produced by strong doses of X rays (Duryee 1939, 1949; von Euler and Hahn 1946). Beyond much doubt, as in the case of ultraviolet radiation (see previous discussion) there is every reason to believe that much of the effect of X rays on cells is due to an interaction of nucleus and cytoplasm, so that those cases in which the irradiated cytoplasm of an egg has no effect on a nucleus introduced later as a result of fertilization (see for example Whiting 1950) do not exclude the possibility that when a nucleus and cytoplasm are irradiated at the same time, the changes in the chromatin may to a large extent originate as a result of some change in the cytoplasm.

Although the cytoplasm of many cells is relatively insensitive to X rays and shows no detectable changes after irradiation, in other cases there is clear evidence that the cytoplasm can be strongly affected by X rays. In their survey of the literature, Heilbrunn and Mazia (1936) listed 24 descriptions of vacuolization in protoplasm exposed either to X rays or to radium. Such vacuolization was found to occur in the cytoplasm of many different types of animal and plant cells. Strangely enough none of this early work is referred to by modern authors. However, reports on vacuolization continue to be published (Gersch and Möbius 1951). The vacuolization reaction is almost always the result of the entrance of free calcium ion into the interior of a cell (Heilbrunn 1928, 1956 a), and it can be assumed therefore that in some cells at least, X rays act much as ultraviolet radiation does. When protoplasm becomes vacuolated, this usually results in a great increase in protoplasmic viscosity. But agents which induce vacuolization in the protoplasm by releasing calcium from the cell cortex generally first cause a decreased viscosity of the protoplasm before vacuolization occurs. This is due to the fact, as will be noted in a later section, that small amounts of calcium when they enter a cell cause a drop in viscosity, whereas larger amounts induce increased viscosity and clotting.

Early studies on plant cells indicate rather clearly that when these cells are exposed either to X rays or to radium, moderate doses cause a decrease in protoplasmic viscosity, whereas larger doses have the opposite effect. This was noted first by LOPRIORE in 1897. He found that a half hour treatment with X rays caused an increase in the rate of protoplasmic streaming in the leaves of the aquatic plant *Vallisneria spiralis,* whereas treatment for an hour caused stoppage of movement and vacuolization of the protoplasm. Similar observations were made by WILLIAMS (1923, 1925) on cells stripped from petioles of *Saxifraga umbrosa,* and exposed either to X rays or radium. Elodea cells showed similar effects when exposed to radium. Of course speed of protoplasmic streaming depends not only on the viscosity of the protoplasm but also on the motive force responsible for the movement. Fortunately WILLIAMS was able to show that the increased speed of streaming was accompanied by increase in the amplitude of the Brownian movement of small granules in the protoplasm. ROCHLIN-GLEICHGEWICHT (1930) likewise showed that radium caused first an increase in the rate of protoplasmic streaming and then after longer or stronger dosage, a decrease. She used cells of the aquatic plant *Elodea* and of the moss *Pterygophyllum.* Her work is especially interesting, for she was able to demonstrate a release of calcium as a result of the irradiation. This was very easy to do, for the cells she studied contain oxalate salts in their vacuoles and when calcium is released into the vacuoles crystals of calcium oxalate appear. FEICHTINGER (1933) studied the effect of alpha, beta and gamma rays on the protoplasm of cells of the alga *Spirogyra.* She used polonium preparations as a source of alpha rays and radium bromide for the beta and gamma rays, and then she made a study of viscosity by centrifuging the algal filaments. In all cases, weak radiation seemed to cause a liquefaction, for the treated filaments had their chloroplasts shifted more readily than did the controls and the opposite effects were obtained with higher doses. The changes in viscosity that FEICHTINGER observed may represent changes in the interior of the protoplasm, the cortical protoplasm, or both. More recently VIRGIN and EHRENBERG (1953), using the centrifuge method, studied the effect of beta and gamma rays on the viscosity of the protoplasm of *Elodea.* They used rays given off by radioactive isotopes. Their results are somewhat confusing. Medium doses seemed to cause a transient decrease in viscosity followed by an increase. Heavier doses seemed to cause only liquefaction.

FORBES and THACHER (1925) exposed eggs of the worm *Nereis* to the beta rays of radium and then after the eggs were fertilized, they centrifuged them along with the control untreated eggs to see if there was any change in the viscosity as a result of the irradiation. Their results indicate that the irradiated protoplasm was somewhat more fluid than the protoplasm of the controls. On the other hand, RIESER (1955 b), studying the effect of strong doses of X rays (50,000 and 100,000 r) on the protoplasm of the eggs of the starfish *Asterias* during the time that the eggs were undergoing maturation divisions, found that the rays caused an increase in the viscosity of the protoplasm. In his experiments, Rieser used a high centrifugal force (22,600 times gravity).

In these studies on *Nereis* and *Asterias* eggs, the viscosity changes occurred during the process of cell division. Eggs not undergoing mitosis do not show any immediate changes in protoplasmic viscosity when subjected to strong doses of X rays, although the same eggs, once they begin mitotic activity, do show such changes. At any rate this is true for the *Arbacia* egg. Thus Wilson (1950) found that when he exposed *Arbacia* eggs to doses of X rays as high as 50,000 r, there was no immediate effect on the

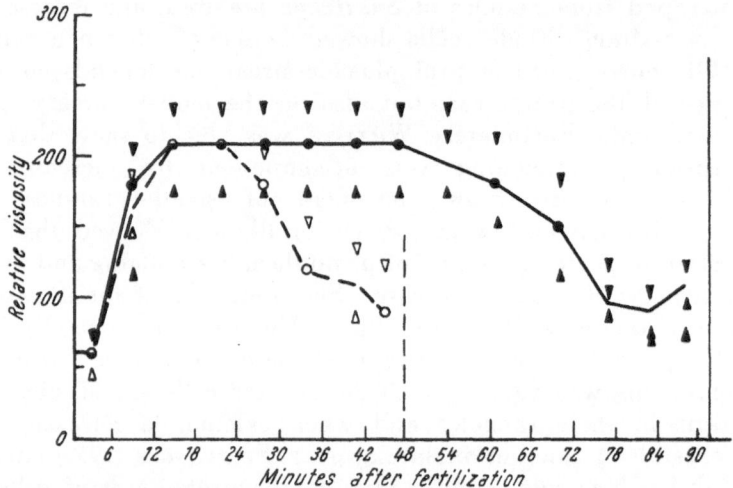

Fig. 13.   The viscosity of the protoplasm of *Arbacia* eggs which have been exposed to 10,000 r and then fertilized, as compared with the viscosity of non-irradiated control eggs. The curve for the controls is a dotted line and the points on it are marked with hollow circles or triangles. The solid line and solid circles and triangles—show the viscosity of the irradiated eggs. A triangle pointed upward indicates that the viscosity is greater than at the point, a triangle pointed downward indicates that the viscosity is lower than at the point.

(Wilson.)

viscosity of the protoplasm. It is only when very much higher doses of X rays are given *Arbacia* eggs that the eggs show a vacuolization indicative of a clotting reaction and an increase in viscosity (unpublished experiment of W. L. Wilson). However, when the eggs are given a dose of only 10,000 r and then fertilized, this relatively small amount of radiation causes marked changes in the viscosity, and cleavage of the eggs is delayed approximately 40 minutes (at 23° C.). In his studies Wilson used the centrifuge method to measure viscosity. His results, or rather the results of one of his experiments, are shown in Fig. 13.

Wilson's work shows clearly that X rays can have effects other than those involving calcium release and consequent changes in the cytoplasm. This is of course to be expected, for modern research has shown that various organic compounds may be affected by X rays. Perhaps in some cells the predominating effect is due to a release of calcium from the cortex. These would be cells in which the combination of calcium in the cortex is sensitive to radiation. In cells not so sensitive it is commonly found that when they are actively in the process of cell division their sensitivity is greatly increased. In such cells, as in cells generally, the protoplasmic colloid undergoes marked cyclic changes in viscosity. These changes appear to be related to the basic colloidal reaction of the protoplasm, a reaction in which a clotting like that of blood occurs (see HEILBRUNN 1956 a). Indeed there is growing evidence that the protoplasmic colloid is in a state of equilibrium between those factors which tend to induce clotting and those which tend to prevent it. In many marine eggs there are metachromatic substances which presumably are akin to heparin (KELLY 1950, 1954). Moreover in the egg of the worm *Chaetopterus,* at approximately the time when the protoplasm becomes fluid after the appearance of the mitotic spindle, metachromatic granules which we call "Kelly granules" because they were first described by KELLY, break down and the metachromatic substance is diffused through the cell (WILSON and HEILBRUNN, unpublished observations). Hence it is logical to assume that the liquefaction of the protoplasm which occurs in the later stages of mitosis and which WILSON found to be retarded or prevented by X rays is dependent on heparin or some similar substance. Thus it could well be concluded that X rays cause a disappearance of heparin from the egg cell, and WILSON in 1950 offered this as a suggestion. This suggestion is now buttressed by the fact that KELLY (1951) was able to show that irradiation may cause a liberation of heparin from combination with protein, and that RIESER and KAYE (1953) were able to demonstrate a release of a heparin-like anticoagulant from irradiated eggs of the clam *Spisula.* RIESER and KAYE used a very high dose of X rays—200,000 r— but much lower doses also cause the release of a metachromatic substance from *Spisula* eggs (unpublished observation of R. O. STEWART).

Our discussion of the effects of various types of radiation on protoplasmic viscosity indicates clearly that the measurement of viscosity provides a sensitive and important index of what is happening to the protoplasm. Recently there has been a great deal of work on the effects of radiation on proteins and enzymes, and although this work has yielded highly significant information as to what X rays do to proteins and enzymes, what we as biologists want to know above all else is how the protoplasm is affected, rather than how some of its constituents react in vitro. And even the meager information that we do have about protoplasm can serve to interpret the action of radiation on certain vital processes. Thus we are in a position to explain why radiation may speed up the streaming of protoplasm, or may slow or inhibit it. We can also understand why ultraviolet radiation can act as a stimulating agent and can induce cells to divide, and from this knowledge it would not be difficult to deduce why it is that X rays can under certain conditions act as a carcinogenic agent. We also have a clue as to why both ultraviolet radiation and X rays can retard cell division. And perhaps some day on the basis of known colloidal changes we may be able to interpret the fact that ultraviolet radiation and X rays can cause mutation. One possible interpretation of this effect is that the release of calcium, which is caused by both types of radiation, could put calcium in contact with chromatin and chromosomes where it might neutralize the negative charge known to be present on the chromatin. Such a neutralization of charge would tend to make the chromosomes stick together, and the resultant adhesions might cause disturbances in the genes. Or if, as Mazia (1954 b) thinks, the chromosome is held together by bivalent cations (Ca or Mg), and if as we have reason to suppose, calcium is set free from its binding with protein both by ultraviolet radiation and by X rays, then such radiation could easily cause chromosome breakage. Ultraviolet might perhaps act more by releasing calcium from the cell cortex, X rays by releasing calcium (or magnesium) from the chromosomes. But as yet these ideas are merely speculative.

The Action of Salts, Acids and Bases

The study of the action of chemical substances on the viscosity of protoplasm is beset with serious difficulties. In the first place, there is always a question as to whether a substance enters into the interior of a cell, and if it does enter, how rapidly it does so. Then too, in so far as the interior protoplasm is concerned, there is the uncertainty as to whether the effects produced there may not be due to some indirect action. For just as temperature changes or irradiation can release the very potent calcium ion from the cortex of the cell, so too it can clearly be shown that many chemical agents have a similar action. In these cases, it is hard to distinguish effects due to calcium released from the cortex and effects due to the substance to which the cell as a whole is exposed. This difficulty was not sufficiently realized in the early work on protoplasmic viscosity, and some of this work needs to be re-evaluated in terms of our present knowledge.

Most cells are relatively impermeable to ions, perhaps not as impermeable as they were once thought to be, but still rather impermeable. Therefore any results obtained with salts or with strong acids or bases must be examined carefully.

There is one type of salt solution whose effect is relatively easy to understand. As is well known, salts of heavy metals coagulate proteins, and thus one might expect them to have a similar effect on protoplasm inasmuch as protoplasm is very largely a protein solution. Heilbrunn (1928) noted that dilute solutions of mercuric chloride or of copper chloride coagulated the protoplasm of eggs of the sea urchin *Arbacia* so that the granules in the protoplasm could no longer be moved by centrifugal force. If the *Arbacia* eggs were first centrifuged and then treated with dilute solutions of mercuric chloride in sea water, new granules could be seen to appear in the hyaline region previously free from granules. Heilbrunn also noted that even with a solution of copper chloride as concentrated as 5×10^{-4} molar, there was a period of time during which the copper did not produce any coagulative effect.

A much more thorough study of the effect of copper on protoplasm was made by ANGERER (1937, 1942). He immersed *Arbacia* eggs in various concentrations of $CuCl_2$. The copper salt was dissolved in solutions which contained either all the common salts of sea water or lacked potassium, magnesium, or calcium. In the absence of calcium, weaker solutions had a more pronounced effect and this effect was more rapidly produced. Apparently the presence of calcium tends to lower permeability to the copper ion. In general all the solutions produced no change in the viscosity of the protoplasm until after the lapse of a time interval which ANGERER

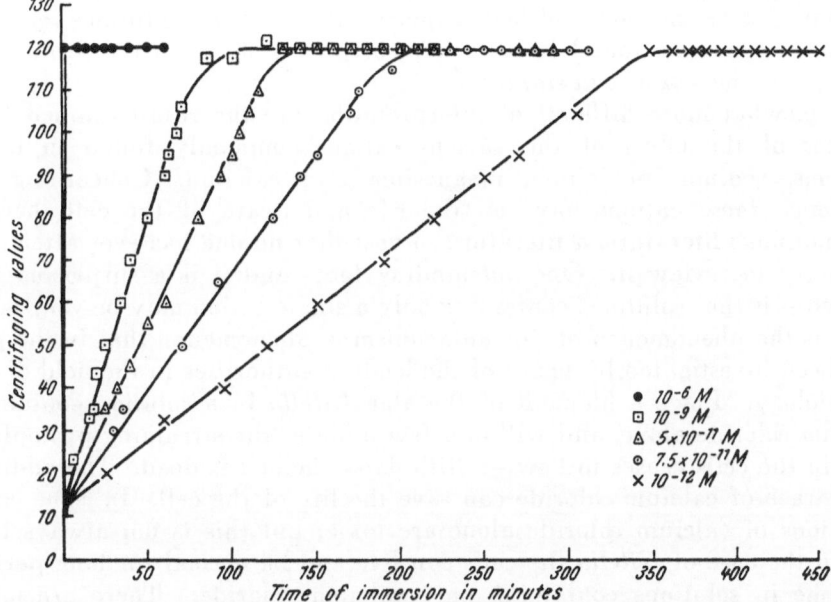

Fig. 14. Change in the viscosity of the protoplasm of *Amoeba dubia* as a function of time after immersion in various dilutions of $CuCl_2$.
(ANGERER.)

calls a latent period. If relatively strong solutions, such as 10^{-3} molar, are used, the viscosity rises to infinity immediately after the expiration of the latent period. But in weaker solutions, as for example in 10^{-4} M $CuCl_2$ in the presence of calcium, after the expiration of the latent period, the viscosity of the protoplasm first decreases until it is only about half as great as it was originally. Then as more copper enters the cell the viscosity of the protoplasm rapidly rises to infinite values. The fact that at first there is a decrease in protoplasmic viscosity is perhaps a phenomenon of the same sort as that which occurs when cells are immersed in solutions of calcium salts. As will be pointed out later, this liquefying effect may be due to an increase in the electric charge of the colloidal particles of the protoplasm.

The protoplasm of ameba is far more sensitive to copper than is the protoplasm of the sea urchin egg. If a copper wire or a copper coin is placed near an ameba, after a short time the ameba dies. A 10^{-6} molar solution has a very rapid effect, and the viscosity of the interior protoplasm increases until it is 10 times its original value. Solutions stronger than 10^{-6} molar are lethal and in such solutions a yellowish brown wave progresses from the periphery of the ameba toward the interior. But solutions of 10^{-6} molar and less do not cause death, and after exposure to these solutions, the ameba can recover. These weaker solutions cause a gelation which is reversible. The course of this gelating or viscosity-increasing effect is shown in Fig. 14. It will be seen from this figure that the viscosity slowly increases with time.

Other heavy metals also cause coagulative changes in protoplasm. For example, in some unpublished experiments, zinc in dilutions as high as 10^{-5} molar was found to have this sort of an effect on the protoplasm of eggs of the worm *Chaetopterus*.

Somewhat more difficult of interpretation are the results gained from studies of the effect of the various cations commonly found in living systems, sodium, potassium, magnesium and calcium. Concerning the influence these cations have on the life and death of the cell there is an enormous literature, a literature so vast that no one has ever attempted properly to review it. One outstanding fact—and it is a surprising fact indeed— is that solutions containing only a single cation may be very toxic. This is the phenomenon of salt antagonism, a phenomenon that in the past has been investigated by many of the leading authorities in the field of cell physiology. Place a filament of the alga *Nitella* in a solution containing sodium chloride alone, and within a few minutes the streaming protoplasm within the cells ceases to flow—a little later the cell is dead. The addition of a trace of calcium chloride can save the life of the cell. In some cases, solutions of calcium chloride alone are toxic; but this is not always true, for in the case of *Nitella* the cells can live and be normal for long periods of time in solutions containing only calcium chloride. There are many instances in the literature like the above.

It is now possible to understand why it is that a solution containing only sodium chloride can be so toxic, for we know now that solutions of sodium chloride can cause calcium to be released from the cortex into the interior protoplasm (for discussion, see Heilbrunn 1952). That this is true is indicated by two types of experiment (among others). The egg of the sea urchin *Arbacia* is relatively impermeable to calcium ions. If it is placed in an isotonic solution of calcium chloride (0.3 molar), calcium enters with extreme slowness. Within the protoplasm of *Arbacia* eggs are red pigment granules that lose their pigment in the presence of minute traces of calcium ion; but only after immersion of the eggs in isotonic calcium chloride solutions for 6 or 7 hours does any breakdown of the pigment granules occur. But if eggs are placed in isotonic solutions of sodium chloride and are washed in these solutions two or three times to remove traces of calcium, then within a few minutes the granules react as if calcium had entered the cell. In order to obtain this result, it is necessary to wash the eggs, for only relatively pure solutions of sodium chloride can act in this way. In such solutions the permeability to sodium ion is increased. And then, apparently, the entering sodium, present in excess in the cortex of of the egg cell, exchanges with the calcium of the calcium proteinate present there, so that free calcium is released into the cell interior. That such a process can actually take place is shown even more strikingly by experiments on *Elodea* cells. These cells conveniently contain soluble oxalate salts in their vacuoles. If *Elodea* leaves are immersed in solutions of calcium salts, little or no calcium enters the cell vacuoles. But on the other hand, as MAZIA and CLARK (1936) showed, immersion of the leaves in pure sodium chloride solutions causes a liberation of calcium into the vacuoles. Some living cells behave in this way and some do not. Skeletal muscle fibers of the frog are impermeable to calcium, but when they are immersed in solutions of potassium chloride they react as though calcium ions were injected into them. On the other hand the muscle fibers of the frog heart appear to be rather freely permeable to calcium and to sodium and potassium as well.

A knowledge of these facts is essential if we are properly to interpret experiments on the effect of various cations on protoplasmic viscosity. HEILBRUNN (1923) was led to investigate the action of various cations on protoplasmic viscosity in order to discover if possible the sign of the charge on the colloidal particles of the protoplasm. Positively charged colloids, especially if they are lyophobic, tend to be precipitated by bivalent or trivalent ions of negative charge, whereas negatively charged colloids tend to be precipitated by positively charged bivalent or trivalent ions. This is a well known phenomenon. Of course the proteins of protoplasm are lyophyllic, but the fact that protoplasm is a suspension and an emulsion would tend to make it behave in lyophobic fashion. At any rate that was the rationale of the experiment. When isotonic solutions of calcium or magnesium chloride were added to the sea water containing *Arbacia* eggs, the calcium and magnesium ions exerted a liquefying influence, that is to say they tended to decrease the viscosity of the protoplasm in the cell interior. On the other hand the addition of varying amounts of isotonic solutions of sodium or potassium chloride to the sea water surrounding the eggs increased the viscosity of the protoplasm. All viscosity tests were made with the centrifuge method. Similar results were obtained when specimens of the ciliate *Stentor coeruleus* were exposed to dilute solutions of sodium, potassium, calcium and magnesium chloride.

Some years later, HEILBRUNN and DAUGHERTY (1931) repeated these experiments with *Amoeba dubia;* they studied the effect of the various cations on the viscosity of the protoplasm in the interior of the ameba. Once again, calcium and magnesium lowered the viscosity, whereas the monovalent ions sodium and potassium had the opposite effect. Consistent results were obtained in a rather large series of tests. More recently, RUNNSTRÖM and KRISZAT (1950) have presented evidence to show that when they added calcium to the sea water containing eggs of the sea urchin *Psammechinus miliaris,* the excess calcium tended to make the protoplasm more fluid. And KRISZAT (1950), on the basis of centrifuge tests, believes that excess calcium causes a decrease in the viscosity of the protoplasm of the giant ameba *Chaos chaos.* (However, as pointed out in the section on

Protozoan Protoplasm, results obtained for *Chaos chaos* with the centrifuge method are difficult to evaluate.)

In judging these experiments of Heilbrunn and others, there is no difficulty in interpreting the action of calcium and magnesium. These ions make the protoplasm more fluid because when they enter the cell they change the ratio of monovalent and bivalent cations so as to add, even though slightly, to the proportion of bivalent cations. This would tend to increase the positive charge on the colloidal particles of the protoplasm, or the positive charge on the suspended particles in the protoplasmic suspension, and it is for that reason presumably that the viscosity is lowered.

But when we come to interpret the action of the monovalent cations, there is certainly difficulty. For the sodium and potassium ions might when they enter the cells, as they presumably do, change the ratio of bivalent to monovalent cations in such a way as to diminish the positive charge on the colloidal or suspended particles of the protoplasm and this would probably tend to increase the protoplasmic viscosity. In the experiments with sea urchin eggs, in which solutions of the salts of the monovalent cations were added to sea water, this may well be all that happens. But when a cell is exposed only to sodium or potassium ions, it is to be expected that in some cases at least calcium is liberated into the interior, and if the amount of such calcium is relatively large, a clotting reaction might be initiated, a reaction which would greatly increase protoplasmic viscosity.

Cholodnyj (1923) immersed root hairs of *Trianea bogotensis* in dilute solutions of sodium chloride or potassium chloride. These solutions caused a progressive slowing of protoplasmic streaming, and then after a time (less than 30 minutes) a protoplasmic clot appeared in the cell and this cloudy white clump of protoplasm was so large that it could be seen with a hand lens. It is possible that in this case the sodium and potassium ions may cause calcium to enter and clot the protoplasm; and yet, according to Cholodnyj, if the root hair with clotted protoplasm is removed from the solution containing only potassium chloride and placed either in a solution of calcium chloride or a solution containing a mixture of potassium and calcium chloride, then after a time there is a liquefaction of the coagulum and a renewal of protoplasmic streaming. Thus although it is possible that the effect of sodium and potassium salts is to cause a release of calcium into the cell interior and that the clotting Cholodnyj observed was due to this calcium, this is not likely, for if it were true, it would be hard to explain the reversal of the clot by the addition of more calcium.

In studies of *Spirogyra orthospira*, Weber (1924 c) found that after the algal filaments had been immersed for long times in solutions of calcium salts, the viscosity of the protoplasm was apparently relatively low. This was indicated by the fact that in such solutions the spiral chloroplasts were more readily moved by centrifugal force than when the algae were in tap water or in solutions of sodium salts.

From these various studies, it is possible to conclude with some certainty that when calcium enters a cell, it can, in small amounts at least, lower the viscosity of the protoplasm in the cell interior. As to the effect of sodium and potassium ions, they increase the viscosity, but this increase may in some instances be due to an indirect action rather than to an action of the ions themselves on the interior protoplasm.

The protoplasm of the cortex shows a very different relation to cations. For the cortex becomes more rigid in the presence of excess calcium and much less rigid in the presence of a increased amount of potassium. This is true for ameba (HEILBRUNN and DAUGHERTY 1932), and it is also true for the *Chaetopterus* egg (WILSON and HEILBRUNN 1952). Indeed the rigidity of the cortex depends on the presence of calcium.

The action of ammonium salts may be similar to that of potassium and sodium salts. This is true for the protoplasm of *Arbacia* eggs (HEILBRUNN 1923) and also for *Trianea* protoplasm (CHOLODNYJ 1923). On the other hand in *Amoeba dubia* HEILBRUNN and DAUGHERTY (1931) found that ammonium chloride caused a decrease in protoplasmic viscosity. This decrease in the case of ameba was accompanied by an alkalinization of the protoplasm, presumably due to the entrance of NH_4OH from the hydrolyzed salt (such alkalinization has been described by JACOBS 1922 b).

If small amounts of calcium tend to make protoplasm more fluid and if this effect is due to an increase of the positive charge on the colloidal and suspended particles in the protoplasm, then it might be expected that trivalent cations would act in the same way as calcium ions only more powerfully. Some years ago, an attempt was made to determine the effect of aluminum ions on the viscosity of the protoplasm of *Arbacia* eggs. There is a serious difficulty in attempting to work with solutions of aluminum chloride, for even rather dilute solutions are strongly acid. If this acidity is neutralized, the aluminum comes out of solution. Thus if one attempts to dissolve aluminum chloride in sea water, the alkalinity of the sea water prevents any of the aluminum staying in solution, unless of course sufficient amounts of the aluminum salt are added to overcome the buffer action of the bicarbonates of the sea water, and then the solutions become strongly acid. Upon neutralization, the aluminum comes out of solution. However by dissolving aluminum chloride in isotonic solutions of sodium chloride—in the absence of calcium—sufficient aluminum remains dissolved when the solutions are brought to neutrality or near neutrality so that the effect of the aluminum ions can be determined. These experiments were never reported in anything but very preliminary form (HEILBRUNN 1925 a). It was found that even in a dilution as great as 4×10^{-6} molar aluminum chloride, the aluminum ion has a marked effect on the protoplasmic viscosity. Indeed it causes a decided lowering of the viscosity. Similar results were obtained with solutions of cerium chloride, likewise prepared in the absence of calcium, but the cerium ion is somewhat less powerful than the aluminum ion. In somewhat higher concentration, the trivalent ions seem to produce secondary effects which result in injury and death of the eggs, and in such solutions the liquefying effect does not occur.

5*

These few experiments with trivalent ions should doubtless be repeated, for the results are of considerable interest and the original study was not sufficiently extensive or complete.

In our search for an understanding of the protoplasmic colloid, it is important to know how it is affected by changes in hydrogen ion concentration. The knowledge that we now have is slight, and the difficulties in the way of obtaining correct information are great. For if we place cells in solutions of varying hydrogen ion concentration, there is no assurance that the hydrogen ion concentration within the cell will be like that of the environment. Indeed, there is every likelihood that it will not be, for in the case of most cells the plasma membrane shows a marked impermeability to hydrogen and hydroxyl ions. Nor is it possible to determine easily or with any degree of certainty the hydrogen ion concentration of the cell interior, and our techniques for making such determinations leave much to be desired (compare Wiercinski 1955).

In studying the effect of acids and alkalies on the viscosity of protoplasm, it is well to remember that in general strong acids and bases are able to enter cells only with great difficulty, whereas weak acids and bases enter cells much more rapidly, for they are only partially ionized and the undissociated molecules can pass through plasma membranes much more freely than hydrogen and hydroxyl ions can. Also it should be remembered that if we compare a strong acid and a weak acid at the same pH, the molar concentration of the weak acid will be much greater than that of the strong acid, and this higher concentration would in itself favor the more rapid penetration of the weak acid into the cell. The same reasoning would also apply to strong and weak bases. Moreover in working with solutions of different pH, it is wise to avoid the use of buffers. Thus, for example, citrates and phosphates are not without effect on protoplasm; both of them have an action on the important calcium ion. Also carbonates and bicarbonates can not be used, for these salts hydrolyze and the carbonic acid enters the cell rapidly whereas the sodium hydroxide does not. Hence in an alkaline solution of sodium bicarbonate the cell interior may become acid (Jacobs 1920). But in the absence of buffers, the pH of a solution may not remain constant. This inconstancy is not a very serious matter if one is working with acid solutions, but alkaline solutions take up carbon dioxide from the air and are constantly changing their pH. To avoid such a change of pH in alkaline solutions it is wise to keep them in an atmosphere free from carbon dioxide, as for example in a desiccator containing solid NaOH. There are various ways in which acids and alkalies can indirectly affect the protoplasm in the cell interior. Both acids and alkalies liquefy the cortex of the *Chaetopterus* egg (Wilson and Heilbrunn 1952) and such liquefaction is commonly associated with a release of calcium into the cell interior. If calcium does enter the cell, this would produce a change in the protoplasmic viscosity. In ameba, the cortex is liquefied by alkalies but not by acids (unpublished observations). If a cell is immersed in an alkaline solution, this would tend to make carbon dioxide leave the cell more rapidly, for as the alkali outside the cell reacted with the carbonic acid, this would favor outward diffusion of carbon dioxide.

Although now and again early students of protoplasm commented on the effect of acid and alkali, there is very little in the way of quantitative information. Some of this older work is discussed critically by HEILBRUNN (1928) and most of it can be discarded as unsatisfactory. VAN HERWERDEN (1925) studied the effect of acetic acid on the leukocytes of human saliva. She immersed them in 0.1% acetic acid in saline solution (0.9% NaCl). Normally these leukocytes show vigorous Brownian movement of the granules in their interior. This movement stopped following exposure to acid, but the effect was reversible and the Brownian movement returned again when the acid was washed off with saline solution. FREDERIKSE (1932) studied the effect of 0.03% acetic acid on the viscosity of the protoplasm of *Amoeba verrucosa*, by observing Brownian movement of granules at the boundary between ectoplasm and endoplasm. The acid caused a sharp increase in viscosity, but after a time— FREDERIKSE doesn't say how long—the viscosity returned to normal values even though the amebae were allowed to remain in the acid solution.

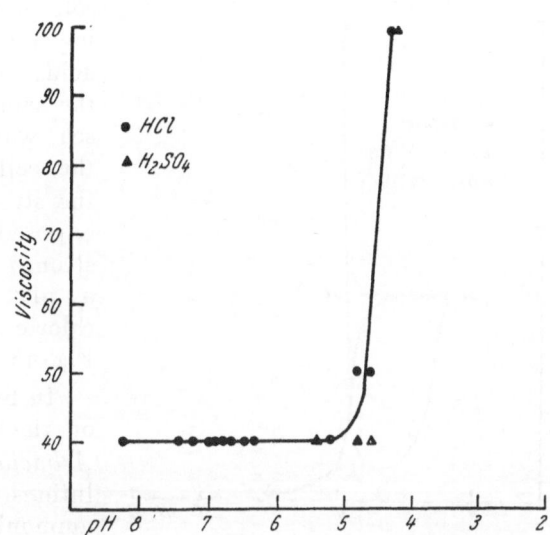

Fig. 15. The effect of HCl and H₂SO₄ on the viscosity of the protoplasm of *Arbacia* eggs. The eggs were exposed for one hour to solutions of varying pH. The viscosity values are in arbitrary units based on times required by the centrifugal treatment to cause movement of granules.
(BARTH.)

JACOBS (1922 a) studied the effect of saturated solutions of carbon dioxide on protoplasmic viscosity. In experiments with *Paramecium caudatum* and *Colpidium colpoda,* he first fed these ciliates with a suspension of India ink so that after some 30 minutes they contained numerous black food vacuoles. Centrifuge tests (at 150 times gravity) showed that the first effect of a saturated solution of CO_2 was to cause a decreased viscosity of the protoplasm. Longer exposures had the opposite effect. Both these effects were reversible, except that after long exposure the protoplasm seemed to be coagulated. When eggs of the sea urchin *Arbacia* were subjected to saturated solutions of CO_2 in sea water, there was a very definite increase in the protoplasmic viscosity as determined by centrifuge tests. This increase also appeared to be reversible. In a few tests, filaments of the alga *Spirogyra* were exposed to saturated solutions of CO_2 and then centrifuged. Here too short exposures to the carbon dioxide seemed to cause a fall in protoplasmic viscosity, whereas longer exposures had the opposite effect.

A much more complete study of the effect of acidity and alkalinity

on protoplasmic viscosity was made by BARTH (1929). He used *Arbacia*
eggs and tested viscosity with the centrifuge methode. When various acids
were dissolved in isotonic NaCl, all of them produced a coagulation of the
protoplasm in 15 minutes when the pH of the solutions was approximately
5.0. The same results were obtained either with weak acids such as carbonic, acetic or lactic; or with strong acids such as hydrochloric or nitric.
However when the acids were dissolved in sea water and the carbon
dioxide resulting from the breakdown of bicarbonates removed, then in
15 minutes the weak acids caused coagulation when the pH of their
solutions was about 5, whereas a much
lower pH was necessary for the strong
acids to produce coagulation within
the same period of time. Evidently in
sea water the weak acids penetrate
the cell much more rapidly than do
the strong acids. When the eggs were
exposed for an hour, both weak and
strong acids caused coagulation at
a pH of 5. The effect of hydrochloric and sulfuric acids is shown in
Fig. 15.

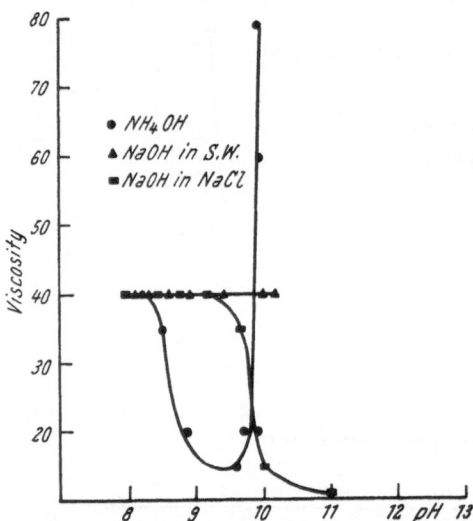

Fig. 16. The action of alkalies on the viscosity
of *Arbacia* egg protoplasm. The circles show
the effect of NH_4OH dissolved in sea water, the
triangles the effect of NaOH in sea water, and
the rectangles the effect of NaOH dissolved in
an isotonic NaCl solution.
(BARTH.)

In his study of the action of alkalies
on the viscosity of the protoplasm of
Arbacia eggs, BARTH found that in solutions of the same pH, the weak base
ammonium hydroxyde had a greater
effect on the viscosity than did the
strong base sodium hydroxide. This
was to be expected, for as is well
known, weak bases penetrate cells
more readily than do strong bases.

Indeed when it is dissolved in sea water, sodium hydroxide has no effect
on the viscosity of the *Arbacia* protoplasm even when it increases the pH
of the sea water from its normal value of 8 to a value of 10.2. On the other
hand, ammonium hydroxide begins to have an effect at pH 8.5. At this
pH it causes a decrease in viscosity, and the decrease becomes more marked
as more ammonium hydroxide is added to the sea water, until finally
a minimum value is reached at pH 9.6. Above this pH the viscosity increases sharply. When sodium hydroxide is dissolved in sodium chloride
solutions instead of sea water, the alkali enters the cell more readily and
causes a marked decrease in viscosity, and this decrease is maintained even
at pH 11. These changes in viscosity are shown in Fig. 16.

BARTH's results indicate that alkalies may have a pronounced liquefying
effect on protoplasm, but that in the case of ammonium hydroxide at least,
higher concentrations of alkali may have the opposite effect. The reason
for this viscosity increase is not clear, but it seems possible that in higher
concentrations alkalies may release relatively large amounts of calcium

from the outer cortical protoplasm. As already mentioned, in both ameba and in *Chaetopterus* eggs, alkalies cause a liquefaction of the cortex, and this may well involve a release of calcium.

HEILBRUNN and WILSON (1955 a) were interested in studying the initiation of cell division in the egg of *Chaetopterus* in an attempt to interpret the action of various parthenogenetic agents in terms of the effect they might have on the protoplasmic viscosity. In the course of this study they found that when HCl was added to sea water until a pH of 4.4 was reached (after the carbon dioxide from the bicarbonates had been allowed to escape), this solution caused an initiation of cell division and also an increase in the viscosity of the protoplasm of most of the eggs. Addition of NaOH in amounts sufficient to cause a rise in the pH of the sea water from 8 to 10, also caused a viscosity increase in the protoplasm of a small percentage of the eggs. This might have been due to the alkali, but it might also have been due to the fact that at pH 10 calcium and magnesium are precipitated out of the sea water.

In general, in spite of the paucity of the data, it can be concluded that acids when they enter the cell cause the viscosity of the protoplasm in the cell interior to increase, whereas alkalies may have a liquefying effect, although in relatively high concentrations the alkalies may also cause the viscosity of the protoplasm to increase. Certainly more information is needed, for it is not at all certain that other types of protoplasm behave like those which have been studied.

Hypertonic and Hypotonic Solutions

The colloidal behavior of protoplasm follows a queer pattern. One surprising and rather striking fact is that agents which might presumably be supposed to act in opposite fashion often enough have the same sort of an effect on protoplasm. So for example, cold and heat both can liquefy, both can coagulate protoplasm, both can act as stimulating agents, and both can anesthetize.

In studies on the artificial parthenogenesis of marine eggs, it has been shown that both hypertonic and hypotonic solutions can cause the eggs to begin their development. HEILBRUNN (1915) found that when *Arbacia* eggs were exposed to a solution consisting of 50 cc. of sea water plus 8 cc. of 2.5 molar NaCl, this hypertonic solution caused a marked increase in the viscosity of the protoplasm as determined by the centrifuge method. Exposure to such hypertonic solutions causes a high percentage of the eggs to cleave and begin their development. Eggs of the worm *Chaetopterus* can also be induced to divide and begin their development as a result of exposure to hypertonic solutions. Indeed in LOEB's original studies of artificial parthenogenesis in *Chaetopterus* eggs he found that treatment with hypertonic solutions gave him the best results (LOEB 1901). HEILBRUNN and WILSON (1955 a) tested the effect of hypertonic solutions on the viscosity of the protoplasm of these eggs. They used the centrifuge method and they found that when the eggs were placed in a solution containing 85 ml. of sea water plus 15 ml. of 2.5 molar NaCl, the protoplasmic viscosity rose

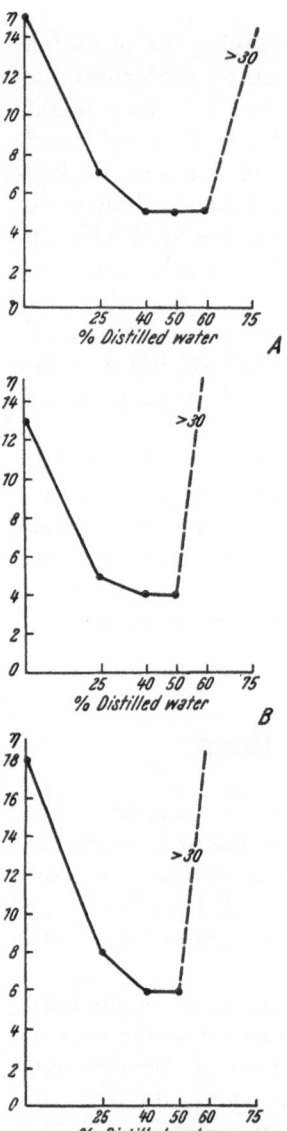

Fig. 17. The effect of hypotonic solutions on the viscosity of the protoplasm of eggs of *Arbacia punctulata*. The 3 curves represent 3 separate experiments. The abscissae give the percentages of distilled water; the ordinates the viscosity in arbitrary units.

in 10 minutes from an arbitrary value of 8 to a value slightly above 60. A little later the viscosity was found to be 70. When the eggs were returned to sea water after varying lengths of exposure (10—80 minutes), the protoplasm soon became very fluid again, the value for the viscosity dropping to about 7.

In the hypertonic solutions, the volume of the cells decreases markedly. This would bring the granules of the protoplasm closer together and might retard their movement when the cells were centrifuged. However, the change in viscosity is so great that it seems certain that the hypertonic solutions do actually cause a viscosity increase. Prolonged exposure to hypertonic solutions causes cytolysis in sea urchin eggs. This reaction, which depends on a vacuolization of the protoplasm, commonly follows excessive treatment with parthenogenetic agents.

When plants are grown under conditions in which moisture is deficient, as might well be expected, the protoplasm shows increased viscosity. This effect has been reported by Schmidt (1939), Schmidt, Diwald, and Stocker (1940) and by Stålfelt (1954). When leaves are placed in distilled water, the viscosity of their protoplasm decreases (Stålfelt 1957).

Hypotonic solutions, or brief exposures to distilled water, can also induce artificial parthenogenesis in sea urchin eggs (for references to the literature, see E. B. Harvey 1956, p. 199). Strongly hypotonic solutions cause an increase in viscosity, an increase presumably due to the cytolytic or vacuolization reaction which occurs both in hypotonic and hypertonic solutions.

In some unpublished experiments, done many years ago, the effect of various hypotonic solutions on the viscosity of the protoplasm of *Arbacia* eggs was determined by the centrifuge method. The results of three of these experiments are shown in Fig. 17. In all of these experiments the eggs were placed in mixtures of distilled water and sea water. In the *A* and *B* experiment, viscosity determinations were made as nearly as possible 15 minutes after entrance of the eggs into the solutions. In the *C* experiment, these determinations were made 30 minutes after immersion. In the *B* experiment the temperature was 25° C., the temperatures at which the other experiments were performed were slightly lower (23.5° for *A*, and 23° for *C*).

In all three tests, the viscosity decreased as the solutions became more dilute until a minimum value was reached; then when 60% or more of the mixture was distilled water, the viscosity rose to infinity, or at least it rose to values above those which could be measured. In such solutions, the protoplasm became cytolyzed, that is to say, it became vacuolated. The fact that distilled water can cause vacuolization of protoplasm has been long known. In early studies of protoplasm, before the osmotic behavior of the cell was understood or appreciated, it was commonly believed that when a cell was placed in a sufficiently dilute solution or in pure water, the entrance of water into the cell was due to the formation of vacuoles in the protoplasm and the increase in size of these vacuoles as more and more water entered. In 1867, the famous botanist HOFMEISTER in his book on the plant cell wrote "Die Entstehung und Ausbildung von Vacuolen ist direct zu beobachten an allen (von der größeren Dichtigkeit der peripherischen Schicht abgesehen) homogenen Protoplasmamassen, welche in Wasser gelangen" and then he goes on to cite various examples. Following HOFMEISTER, and probably also before him, there have been various descriptions of the vacuolization of protoplasm following the rapid entrance of water into cells. Such a vacuolization would have a marked effect on the physical properties of the protoplasm. Certainly if a cell becomes filled with vacuoles it can no longer behave as a fluid. HOFMEISTER himself believed that the entrance of excess water into a cell could cause coagulation.

Fat Solvents

All types of living material are sensitive to fat solvents, even when these are present in low concentration. When excitable cells or tissues are exposed to a dilute solutions of ether, they typically lose their ability to respond to stimulation; that is to say they become anesthetized, or as most pharmacologists would prefer to say, narcotized. Physiologists and pharmacologists have long sought the reason for this anesthetic action. As was first recognized by CLAUDE BERNARD, and recognized clearly, the answer is to be found in a study of the cell and its protoplasm. The subject of anesthesia and the effect of anesthetics on protoplasm will be discussed in a special section of this treatise. In that section it will be pointed out that most of the earlier attempts to explain anesthesia in terms of the physiology of the cell have not been successful. Anesthesia is certainly not universally due to a decrease in the permeability of the plasma membrane, nor can it any longer with justice be held that the primary action of fat solvents is to decrease the activity of oxidizing enzymes. But anesthetics of all sorts do affect the viscosity of the protoplasm, and this is the only significant effect that they are known to have on the chemical or physical properties of the protoplasm.

In considering the effect of a substance like ether on protoplasmic viscosity, we do not need to be worried as to whether or not the substance enters the cell interior. For fat solvents of all sorts are known to enter cells with extreme readiness. Nevertheless we can not exclude the possibility that the fat solvents might have some indirect effect on the interior protoplasm. For if the cortex of a cell contains bound calcium which can be freed by various agents, then it is of interest to inquire as to whether or not fat sovents might not act in this way. Actually there is evidence that they do. In the first place, solutions of ether and other fat solvents certainly cause a liquefaction of the cortex both in the ameba (Daugherty 1937) and in the *Chaetopterus* egg (Wilson and Heilbrunn 1952), and such liquefaction is generally associated with calcium release. By chemical analysis, Berwick (1951) showed that ether and chloroform released calcium from frog muscle fibers. And finally Mazia and Clark (1936) were able to show that when cells of the water plant *Elodea* were exposed to solutions of ether, calcium was released from the protoplasm and caused a precipitation of calcium oxalate in the vacuoles.

The conclusion to be drawn—and it undoubtedly is the correct conclusion—is that solutions of fat solvents may have a double effect on the protoplasm of the cell interior. One effect would be due to the direct action of the fat solvent itself; the other effect would be a consequence of the entrance into the cell of calcium ion released from the cortex. If we keep this point of view in mind, we will be able properly to interpret the strange fact that solutions of a fat solvent, differing only slightly in concentration, can have quite opposite effects on the protoplasmic viscosity. This is true not merely for one type of cell but for all types of cells that have been carefully investigated. In thinking in terms of two opposing actions, we are in a position to explain the puzzling fact that fat solvents may under one set of conditions act as anesthetic agents and under a not very different set of conditions as stimulating agents. As a matter of fact there is also a third possible effect that a solution of a fat solvent may have on a cell, especially an isolated cell like a marine egg cell or a blood cell. In relatively high concentrations of ether in sea water, sea urchin eggs tend to swell. This swelling might be due to a lowering of interfacial tension at the outer surface of the cell, as well as a lowering of interfacial tension at the boundaries of the cytoplasmic granules and vacuoles; it might also be due to a marked increase in the permeability of the plasma membrane.

Before anyone was able to make any quantitative measurement of protoplasmic viscosity, students of protoplasmic streaming noticed that dilute solutions of fat solvents caused the protoplasm to stream more rapidly, whereas more concentrated solutions had the opposite effect. Such observations were made by Klemm (1895), Josing (1901), and especially by Ewart (1903). These effects on streaming might be due to changes in protoplasmic viscosity, but they might also be due to changes in the motive force responsible for the streaming.

HEILBRONN (1914) was the first to make measurements of the viscosity of the protoplasm of cells exposed to ether solutions. He placed starch sheath cells of bean plants in various concentrations of ether in water and measured protoplasmic viscosity by observing the speed of fall of the starch grains. The concentrations of ether that HEILBRONN used are not clearly stated; he used various percentages of a saturated aqueous solution. What he found was that dilute solutions caused a drop in viscosity; more concentrated solutions (approximately 1.5% by volume) caused a reversible gelation, and still more concentrated solutions a gelation that was irreversible. HEILBRONN believed that the 1.5% solution was anesthetic and that it would prevent geotropic response. As a matter of fact, in most plant material it is not easy to decide what constitutes anesthesia, and certainly response of a plant to the stimulus of gravity is rather different from the response of a muscle or a nerve. It may well depend on the movement of starch grains acting as statoliths, and such movement through the protoplasm would be prevented by gelation.

WEBER (1922) repeated HEILBRONN's work on the effect of ether on the viscosity of the protoplasm of bean cells. He used starch sheath cells in the epicotyls of *Phaseolus vulgaris,* and determined viscosity by the centrifuge method. Exposure to 2½ and 5% ether (by volume) caused a sharp increase in protoplasmic viscosity, an increase which was estimated to be about 12-fold. This viscosity increase was reversible. A 10% solution of ether (by volume) caused an irreversible increase in viscosity, and death resulted.

WEBER (1921) used the centrifuge method in a study of the effect of ether solutions on the viscosity of the protoplasm of the alga *Spirogyra.* Following exposure to relatively low concentrations of ether, that is to say to concentrations of 1–2½ volume p. c., there was a decrease in viscosity when the algal filaments were exposed for 1–2 hours. Higher concentrations (about 3% or more) had the opposite effect and the chloroplasts were less readily moved by centrifugal force. At an intermediate concentration of 2.5–3%, the effect varied with the length of exposure. An exposure of one hour caused a decrease in viscosity, whereas an exposure of 3 or more hours had the opposite effect. In the discussion of his experiments, WEBER realized that the displacement of the spiral chloroplasts by centrifugal force may not be a true indication of the viscosity of the interior protoplasm, for there is some uncertainty as to the way these chloroplasts are attached to the periphery of the cell. He pointed out that in the weaker solutions of ether, the protoplasmic streaming was more active, whereas in a 3% solution of ether, the protoplasmic streaming was slowed or stopped. WEBER in this early work found that the effects of both low and relatively high concentrations of ether were reversible.

In 1925, WEBER (1925 a) used his plasmolysis method of studying viscosity to determine the effect of dilute (2%) solutions of ether on protoplasmic viscosity. When normal cells of *Spirogyra varians* were plas-molyzed by being placed in 20% sucrose, the contour of the plasmolyzed cells was angular, indicating high viscosity; whereas etherized cells sub-jected to the same treatment showed what Weber calls "convex" plas-molysis, indicating low viscosity. This is shown in Fig. 18. Weber thought the effect of the ether to be due to a change in the viscosity (and the adhesiveness) of the protoplasm in the "Hautschicht," that is to say, the cortex. Similar results with the plasmolysis method were obtained for cells of the aquatic plant *Elodea* and for cells of the stalk of *Callisia repens,* a plant of the family Commelinaceae. In this 1925 paper, Weber discusses again his earlier centrifuge experiments with *Spirogyra* cells, and he concludes that inasmuch as the chloroplasts after breaking loose are moved through the interior protoplasm, his results indi-cate that 2% ether causes a lowering of viscosity both of the cortex and of the interior protoplasm or endoplasm. In the 1925 paper Weber finds that 3% ether causes death of most of the cells after an ex-posure of 40 minutes and he states that the reversible, viscosity-lowering effect of the 2% ether indi-cates that this represents the narcotic concentration.

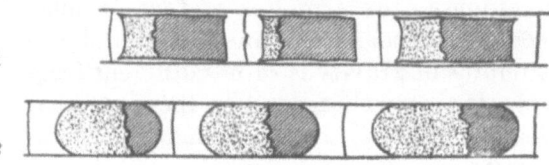

A

B

Fig. 18. *A* Angular plasmolysis in normal cells of *Spirogyra varians*. *B* Convex plasmolysis in cells previously exposed to 2% ether.
(Weber 1925.)

Northen (1938 b, 1939) also studied the effect of ether on the protoplasm of *Spirogyra*. He also, at least for higher concentrations of ether, found a significant change in the ease with which the chloroplasts moved under the influence of centrifugal force, and he concludes that ether—and other fat solvents as well—cause a liquefaction of the protoplasm. But he is apparently not as aware as Weber of the fact that ether may be acting on both the cortical protoplasm and the interior protoplasm. Presumably, at least in the light of work on other types of protoplasm, both the outer and the inner protoplasm may be affected.

Kessler (1935) studied the effect of chloroform vapor on the proto-plasmic viscosity of leaves of the plant *Sempervivum glaucum*. Both with the centrifuge method and with Weber's plasmolysis method, he found that the chloroform lowered the viscosity. But in this case also one can not be sure as to whether Kessler was measuring the viscosity of the cortical protoplasm, the interior protoplasm, or both.

One advantage of marine eggs for the study of protoplasmic viscosity is that one can easily distinguish between the cortex and the interior. Heilbrunn (1917, 1920 a, 1920 b) studied the effect of ether and other fat solvents on the viscosity of the interior protoplasm of fertilized sea urchin (*Arbacia*) eggs. After it is fertilized, the sea urchin egg begins to prepare for cell division. In the course of this process the viscosity of the proto-plasm in the interior of the cell increases to three or four times its original

value (see Fig. 13). However if the cells are placed in a solution of 2½%
ether (by volume), no such increase in viscosity occurs and the mitotic
spindle does not form. Thus ether prevents the gelation which normally
occurs following fertilization. Various other fat solvents also prevent the
normal mitotic gelation; among others, chloroform, acetone, nitromethane,
ethyl acetate, propyl alcohol and amyl alcohol. This prevention of gelation
only occurs when the fat solvents are present in low concentration. Higher
concentrations cause coagulation and death of the cells. Not only does
ether prevent the gelation in fertilized sea urchin eggs, it also tends to lower
the viscosity of the protoplasm in unfertilized eggs. HEILBRUNN (1925 b)
found that a 2½% solution caused a decided decrease in the viscosity of
the protoplasm of these eggs as determined by centrifuge tests. On the
average, the viscosity dropped to about one half of its original value.

In sea urchin eggs it is not possible to make proper measurements of the
viscosity of the cortical protoplasm, for the small granules of the cortex
do not shift readily under the influence of centrifugal force. However in
the egg of the worm *Chaetopterus*, in which the cortical granules are larger
and the cortex perhaps less rigid, it can readily be shown that dilute
solutions of ether and other fat solvents tend to liquefy the cortex. According to WILSON and HEILBRUNN (1952), when *Chaetopterus* eggs were exposed to a 2% ether solution, the centrifugal force required to dislodge the
granules from the cortex was little more than a fourth of that required to
dislodge the granules from the cortex of the control, untreated eggs.
A similar liquefying effect was produced by 1.5% butyl alcohol and 0.5%
amyl alcohol.

Thus for various types of protoplasm, fat solvents may have a liquefying
action both on the cortex and on the interior protoplasm. The liquefaction
of the cortex almost certainly is associated with a release of calcium and
this calcium entering the cell can have a clotting or coagulating effect.
If sea urchin eggs are exposed to relatively high concentrations of ether, for
example to a 3.5% solution, then the liquefying effect of the ether is over-
ridden by the relatively large amount of calcium that has entered the egg
interior. The protoplasm in the cell interior becomes cytolyzed or
vacuolated just as it always does in the presence of an excess of calcium.
This vacuolization or clotting reaction is a reaction similar to that which
occurs when a cell is torn or broken, the surface precipitation reaction or
s. p. r. In the sea urchin egg, ether can prevent the s. p. r. only if a
relatively small amount of calcium is present; if larger amounts of calcium
are present, then the clotting reaction can occur in spite of the ether; HEIL-
BRUNN (1934). Typically in the living cell there is very little free calcium
ion. That is why dilute solutions of ether can exert a liquefying or anticlotting effect on the protoplasm of the cell interior.

Several workers have studied the effect of ether on the viscosity of the
protoplasm of ameba. BRINLEY (1928) from observations on the rate of
Brownian movement in amebae exposed to solutions of ether, chloroform
and alcohol, concluded that these fat solvents caused a decrease in the

viscosity of the protoplasm. An attempt at a more quantitative study was made by Frederikse (1933). He experimented on *Amoeba verrucosa*, an ameba which apparently is tightly packed with granular inclusions. Frederikse used the Pekarek method of measuring the viscosity, a method involving the observation of Brownian movement. This method, as pointed out earlier, can not safely be used if the concentration of granular material is high, as apparently it is in *Amoeba verrucosa*. As might be expected, Frederikse's data show a high degree of variability, but they do indicate that dilute solutions of chloroform lower the viscosity of the protoplasm in the interior of the ameba.

No doubt the most careful and the most dependable work on the effect of fat solvents on the viscosity of ameba protoplasm is that of Daugherty (1937). Some of her work was also reported by Heilbrunn (1951, 1956 a). In her experiments she used the centrifuge method, and she measured the viscosity of the cortical protoplasm by subjecting *Amoeba proteus* to rather strong centrifugal forces. Her studies on the protoplasm in the cell interior were made on *Amoeba dubia* and in this case much lower centrifugal forces were used. Daugherty's studies show clearly that the cortical protoplasm of the ameba undergoes a decrease in viscosity when it is exposed to dilute solutions of fat solvents. Thus 2 molar methyl alcohol causes a drop in the viscosity of nearly 50% and a similar effect is produced by a molar solution of ethyl alcohol. Still weaker solutions of higher alcohols act in the same way; a somewhat lesser decrease in viscosity is produced by 0.2 molar ether (2% by volume).

In so far as the interior protoplasm is concerned, Daugherty found that dilute solutions of fat solvents could cause a pronounced drop in the viscosity. This is shown in the following table, condensed from Daugherty's data.

The effect of fat solvents on the viscosity of protoplasm in the interior of Amoeba dubia

Reagent	Time of exposure minutes	Viscosity decrease
1.5% *n*-butyl alcohol	20–29	50%
0.7% iso amyl alcohol	15–23	62%
2% ether	18–28	50%

Thus dilute solutions of fat solvents can markedly lower the viscosity of the interior protoplasm. But inasmuch as these solutions release calcium from the cortex, it might be supposed that if just enough calcium were released, under certain conditions the viscosity lowering effect of the fat solvent would be more than offset by the clotting effect of the calcium. And this is indeed true. For very dilute solutions of butyl alcohol can act in opposite fashion from slightly more concentrated solutions. This is shown in the following table.

The effect of various concentrations of butyl alcohol on the viscosity of the interior protoplasm of Amoeba dubia

Concentration	Normal viscosity	Time of exposure minutes	Viscosity after exposure
0.5%	6	26	45
1.0%	5	6	5
		18	10
		56	40
1.25%	8	8	3
		17	2

Entirely similar results were obtained with various concentrations of ether, with 1% solutions increasing the viscosity by 100–200% and 2% solutions decreasing the viscosity to half of its original value. These results could serve to explain why it is that in many instances dilute solutions of fat solvent anesthetics have an effect quite opposite to that produced by solutions of somewhat higher concentration. As a matter of fact, in an ameba it is hard to decide just what to call anesthesia or narcosis. An ameba moves most of the time—stimulating agents cause it to stop, just as stimulating agents always cause a cessation of the protoplasmic streaming of plant cells. DAUGHERTY found that when amebae were exposed to ultraviolet radiation, which is a wellnigh universal stimulating agent, and one which normally causes viscosity increase in the interior of the ameba, the presence of 1.5% butyl alcohol or 2% ether would prevent the viscosity-increasing action of the stimulating agent. Hence these are solutions which can best be considered to have a narcotic action.

More work needs to be done on the effects of various fat solvents on protoplasmic viscosity, but what has been done gives a clear indication that the protoplasmic colloid is sensitive to dilute solutions of these substances and reacts to them in ways that are easily detectable by viscosity measurements.

Cyanide and Azide

Both cyanide and azide have a powerful effect in blocking the cytochrome oxidase system in living cells, and it has always been assumed that the relation of vital processes to these substances is entirely due to their action on one of the chief oxidation systems of protoplasm. It is of interest therefore to inquire as to whether or not cyanide and azide have an effect on the viscosity of protoplasm. For in the living colloid, the ever-present oxidative reactions must in some way or another be related to changes in the colloidal state. Indeed one of the major problems the cell physiologist has to face is the question as to how the metabolism of the cell is related to its physical activity. For example we very much need to know why it is that when a muscle is thrown into activity and made to contract, its oxidative processes increase in intensity.

How does cyanide affect the protoplasmic colloid? In the first place it should be emphasized that cells vary widely in their sensitivity to cyanide. This was clearly pointed out by Robbie (1949). In his study of the effect of cyanide on the respiration of various invertebrate tissues, he found that these tissues showed a thousand-fold variation in their sensitivity to cyanide. This may in large measure be due to variations in the cellular respiration systems in the different organisms. But in any attempt to interpret vital phenomena it should be borne in mind that in living cells oxidative and proteolytic enzymes are not sharply separated from each other, and that they may strongly influence each other. Thus proteolytic enzymes could easily digest and destroy the oxidizing enzymes and oxidation of sulf-hydryl groups might impair the activity of proteolytic enzymes (see for example, Bersin 1935, Hellerman 1937), or it might also have an effect on the activators or the inhibitors of these proteolytic enzymes. The reason that this has an importance for our understanding of the protoplasmic colloid and its behavior as indicated by viscosity changes is the fact that, as we shall see more fully later, the main reason for sharp increases in protoplasmic viscosity is a clotting reaction due essentially to the action of clotting enzymes which are at the same time proteolytic enzymes.

Studies have been made of the effect of cyanide on the protoplasmic viscosity of three types of marine eggs. These eggs vary widely in their sensitivity to cyanide. Thus the egg of the sea urchin *Arbacia* is 500 to 1,000 times more sensitive than is the egg of the worm *Chaetopterus*. The egg of the clam *Spisula* occupies an intermediate position. In all three of these eggs, solutions of potassium cyanide prevent cell division, and they do this without killing the eggs. According to Wilson and Heilbrunn (1957), when fertilized *Chaetopterus* eggs are placed in a 10^{-2} molar KCN solution (adjusted to pH 8), the gelation which normally precedes the formation of the mitotic spindle is completely inhibited, and the protoplasm remains fluid. Likewise in the *Spisula* egg dilute solutions of KCN keep the proto-plasm fluid and prevent the mitotic gelation. The reason for this may be the fact that in the mitotic gelation, as in protoplasmic clotting generally, there is an oxidative change in which sulfhydryl groups are oxidized to disulfide groups. If this oxidative change is inhibited, the protoplasm can not clot. This point of view is supported by the fact that parachloro-mercuribenzoic acid in concentrations of about 10^{-4} molar also prevents the mitotic gelation. Parachloromercuribenzoic acid inhibits the oxidation of the SH group, an oxidation which normally occurs during the formation of the mitotic spindle (Rapkine 1931, Mazia 1954 a).

But whereas cyanide prevents the mitotic gelation in the eggs of *Chaetopterus*, it was found that it had a very different effect on *Arbacia* eggs. In these eggs, a very dilute solution of KCN (10^{-5} molar) enhances the mitotic gelation and prevents the reversal of it, a reversal that normally occurs in the course of the mitotic process. In other words, in the presence of cyanide, the viscosity of *Arbacia* egg protoplasm, once it rises in preparation for the formation of the mitotic spindle, stays high and does not drop again after the spindle has formed. It is hard to understand why cyanide should act in one way on some types of protoplasm and in exactly the opposite way on other types of protoplasm. The suggestion has been made (HEILBRUNN 1956 a) that cyanide acts not only on respiratory enzymes but on proteolytic enzymes of the cell as well. It is well known that cyanide favors the action of some proteolytic enzymes (PURR 1935), but that it tends strongly to inhibit the action of others (SUGAI 1944; WILLSTÄTTER, GRASSMANN and AMBROS 1926). In those instances in which cyanide favors the action of proteolytic enzymes, as it does in the case of cathepsin and papain, protoplasmic clotting would be favored; for as will be noted later, proteolytic enzymes cause clotting both in blood and in protoplasm. This would explain the action of cyanide on the protoplasm of the *Arbacia* egg. On the other hand, when cyanide inhibits the action of proteolytic enzymes, as it does in the case of trypsin and some plant proteases, then it would tend to prevent protoplasmic clotting. This might be an important factor in determining the action of cyanide on the protoplasm of the *Chaetopterus* and the *Spisula* egg.

Although azide, like cyanide, inhibits the activity of the cytochrome oxidase system, it is much less potent than cyanide in its antimitotic effect on *Arbacia* eggs. In order to stop cell division in this egg, approximately a 10^{-2} molar solution of sodium azide must be used; this is about a thousand times as high a concentration as is required for potassium cyanide. When fertilized *Arbacia* eggs are placed in a 10^{-2} M solution of sodium azide two minutes after fertilization, the mitotic gelation is delayed and the protoplasm of the treated eggs may still retain its original fluidity while the normal eggs are beginning to show a mitotic gelation. But later when the protoplasm in the normal eggs has become more fluid again, the treated eggs show a relatively viscous protoplasm. Thus on *Arbacia* eggs azide seems to have a double effect, as might perhaps be expected from the earlier discussion. On removal from the azide solution, the eggs recover and divide normally. In the *Spisula* egg, although the data are still rather meager, it seems that 10^{-2} M sodium azide acts much like 10^{-4} M potassium cyanide in preventing the mitotic gelation.

Clotting and Anticlotting Substances

If it is true, as has already been urged, that increases in protoplasmic viscosity are in large measure due to a clotting reaction similar in many respects to the clotting reaction that occurs in vertebrate blood, then it might be expected that substances which clot vertebrate blood would act in the same way on protoplasm, and substances which prevent the clotting of vertebrate blood would tend to keep the protoplasm fluid. In vertebrate blood, as is now well known, clotting is due to the action of a calcium-activated proteolytic enzyme, thrombin, on the protein fibrinogen. Other proteolytic enzymes, such as trypsin, can also cause blood to clot. (There are many recent reviews on various phases of the blood clotting process; see for example Lorand 1954.)

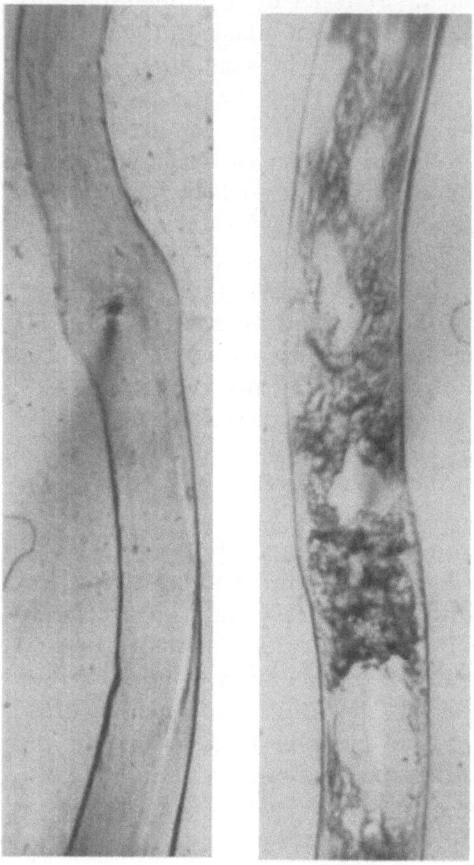

Trypsin has a marked clotting effect on the protoplasm of frog muscle fibers (Woodward 1948, Wiercinski and Cookson 1949). Wiercinski took a motion picture of what happens when a dilute solution of trypsin is injected into the interior of a frog muscle fiber. Two of the frames from this motion picture are shown in Fig. 19. Even without viscosity measurements, there can scarcely be any doubt that the trypsin solution caused a violent coagulation of the protoplasm with an increase in the viscosity to infinity. Similar effects, though not so violent, could be obtained with much more dilute solutions of the enzyme.

The clotting of vertebrate blood can be inhibited either by the removal of ionized calcium or by the addition of the mucopolysaccharide heparin, a substance which not only inhibits blood clotting but also retards or inhibits the action of various proteases. In some types of cells, heparin also prevents the clotting reaction which occurs when a cell is torn or broken; that is to say, it inhibits the surface precipitation reaction. Unpublished

Fig. 19. The action of a dilute solution of trypsin (0.05%) on the protoplasm of a frog muscle fiber. The photograph at the left shows the entrance of the micropipette containing the trypsin solution. The photograph at the right shows the result of the injection.
(Wiercinski.)

observations show clearly that dilute solutions of heparin inhibit the surface precipitation reaction in the giant ameba *Chaos chaos*. This would indicate that heparin might prevent protoplasmic clotting within cells and this indeed is the case. Thus Heilbrunn and Wilson (1949) found that when eggs of the worm *Chaetopterus* were exposed to 0.05% heparin and then fertilized, the viscosity increase that normally precedes the formation of the mitotic spindle did not occur; instead the protoplasm remained fluid. Perhaps heparin in most instances can prevent clotting only when the concentration of calcium ion is low. This is indicated by the fact that when

Arbacia eggs are broken in sea water containing heparin, the surface precipitation reaction is not prevented.

A substance which has a very potent effect in slowing or preventing the clotting of human blood is dicoumarol. This substance, originally prepared from clover which had spoiled, is a structural analogue of vitamin K, and like the various kinds of vitamin K is a naphthoquinone derivative. Both vitamin K and dicoumarol have an indirect effect on the coagulation of blood. That is to say, they do not exert much of any action when added to blood directly, but only when given to the whole organism, and then only after a lapse of time. Both vitamin K and dicoumarol have strong and striking effects on protoplasmic viscosity, and they act in exactly opposite fashion. First as to dicoumarol. Its action was tested on the fertilized eggs of *Arbacia* and *Chaetopterus* (HEILBRUNN and WILSON 1950). When dicoumarol is added to sea water, there is a rather voluminous precipitate, and apparently dicoumarol forms insoluble compounds with the

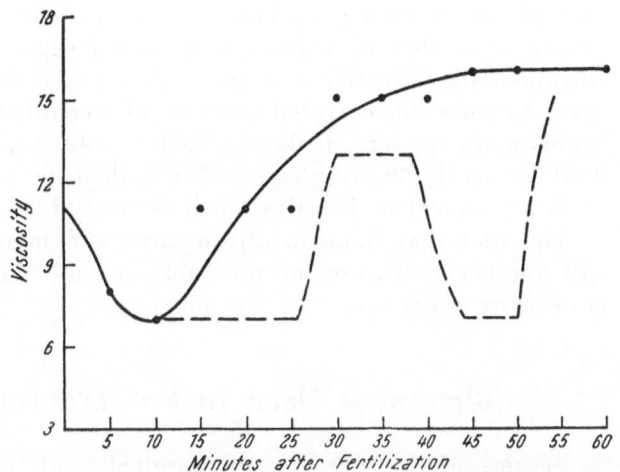

Fig. 20. Effect of a dilute dicoumarol solution on the protoplasmic viscosity of *Chaetopterus* eggs placed in the solution 2 minutes after insemination and kept at 21° C. The broken line shows the viscosity of normal eggs for comparison; this curve is taken from HEILBRUNN and WILSON (1948). Viscosity values for both curves are in arbitrary units.

salts of sea water. Hence the concentrations used were rather uncertain, and all that can be said is that a concentration of something less than 0.01% was the one used in the experiments. When *Chaetopterus* eggs were fertilized and then put in the solution of dicoumarol two minutes later, the protoplasm of these eggs instead of going through the usual cycle of viscosity changes, shows instead a decided increase in protoplasmic viscosity, an increase which is maintained for several hours. The values of the viscosity for the first hour after fertilization are shown in Fig. 20. Subsequently, the viscosity of the protoplasm begins to drop. This drop in viscosity does not occur uniformly in all eggs, so that it is scarcely possible to plot a curve, but it is clear from the data that after about 5 or 6 hours at least half of the eggs have protoplasm with a viscosity below what is normal either for unfertilized eggs or for fertilized eggs when the viscosity is lowest. When the protoplasm becomes markedly fluid, the eggs divide, long hours after the control eggs have divided. They make a single division and then stop. Thus dicoumarol produces liquefaction of the protoplasm, but only after a long time interval.

Vitamin K acts in exactly opposite fashion. One of the most potent of the vitamin K compounds is 2-methyl-1, 4-naphthoquinone, sometimes called menadione. This substance has the advantage of being soluble in water, although it is only sparingly soluble. In sea water a solution slightly less concentrated than 0.001 % can be obtained. Such a solution has a powerful effect on *Chaetopterus* eggs; it completely stops division. Indeed a solution containing only 0.0001 % was also effective in stopping division in over 90 % of the eggs. Such a solution is 5.6×10^{-6} molar. When *Chaetopterus* eggs, two minutes after fertilization, were placed in this solution, the mitotic gelation was inhibited and the protoplasm remained fluid for hours. Then five hours after fertilization, some of the eggs began to show an increased protoplasmic viscosity, an hour later 70 % of the eggs had protoplasm more viscous than that of normal unfertilized eggs, and at 7 hours after fertilization, the viscosity was more than twice that of normal unfertilized eggs. In more concentrated solutions of menadione, the increase in viscosity begins much earlier. Thus in a 0.001 % solution, the fertilized eggs remain fluid for nearly 50 minutes (at 21^0 C.), then the viscosity rises rapidly until at 76 minutes after fertilization it is very high.

The fact that heparin, dicoumarol and menadione act as they do is still another indication of the fact the colloidal behavior of protoplasm is in many ways very like that of blood.

Substances Used in the Treatment of Cancer

Because of the fact that the surgical treatment of cancer has not been overly successful even when combined with radiation therapy, workers in various parts of the civilized world have been eagerly searching for some chemical remedy for this dread disease. Great screening programs have been established, and thousands of compounds have been prepared and tested at industrial establishments, at universities, and at giant research institutes. For the most part, the search for suitable compounds has not been conducted on any very rational basis, and the leads for experimentation have mostly come as a result of lucky accidents. One of the leading workers in the field, Haddow, in writing about substances which cause cancer and those which tend to cure it, wrote in 1951: "But in no case—a striking fact—do we know the place in the cell at which they act—whether the cell surface, the cytoplasm, the nucleus itself- -or the nature of the receptors with which they combine." Whatever efforts have been made toward an interpretation of the action of chemotherapeutic agents for cancer have mostly been centered around the idea that these agents cause some disturbance of metabolism. But some of the most powerful chemotherapeutic agents do not affect the growth of the cell. Thus in the presence of nitrogen mustard, or its oxygen derivative, Nitromin, cells increase in size but do not divide (Bodenstein 1947; Friedenwald, Buschke and Scholz 1948; Sato, Belkin, and Essner 1956).

To add to the difficulties in the way of interpretation is the strange and puzzling fact that the very agents which have been found most effective in treating cancer are not infrequently capable of acting as carcinogenic agents. This was first pointed out by HADDOW (1935), and he and his collaborators have written a number of papers on the subject. In his excellent book on the "Biochemistry of Cancer," GREENSTEIN (1954) frequently refers to this phenomenon as the "Haddow paradox." But neither HADDOW nor GREENSTEIN offers any explanation of the paradox, and all that the latter offers as a suggestion is the wry remark that "The capacity for intellectual flexibility combined with scientific care is one of the demands in the field."

There is however, on the basis of our knowledge of viscosity changes, a simple interpretation of the action of various therapeutic agents that have been used in the treatment of cancer, and it is also possible on the basis of viscosity changes to explain without much difficulty the mysterious HADDOW paradox.

One of the most interesting substances that has been used in the treatment of cancer is the substance colchicine ($C_{22}H_{25}O_6N$) generally regarded as an alkaloid. So much work has been done on this substance by agriculturists, biologists and clinicians that an entire book has been written about it (EIGSTI and DUSTIN 1955), and this may be consulted for references to the literature. Long used as a cure for gout, physicians noted that colchicine had a favorable effect on cancer. But it was not until 1934 that it was clearly recognized as an agent which had an effect on mitosis and could retard the progress of a neoplasm. Unfortunately, however, colchicine is a rather toxic substance so that it can not very well be used in the treatment of human cancer. Our interest is in why it has any effect at all. Beyond any doubt dilute solutions of colchicine prevent cell division. Indeed much has been written about the effect of colchicine on mitosis and this literature has been reviewed by EIGSTI and DUSTIN. If fertilized *Arbacia* eggs are placed in dilute solutions of colchicine (2×10^{-5} M), the normal mitotic gelation is inhibited (BEAMS and EVANS 1940) and cell division does not occur. This was also shown to be true by WILBUR (1940); he made many determinations of the viscosity. WILBUR used somewhat more concentrated solutions than did BEAMS and EVANS—also he studied the effect of these solutions on unfertilized eggs. In these unfertilized eggs, the colchicine has no effect on the viscosity. Apparently colchicine has some effect on the protoplasmic clotting system. In this connection, it is interesting to note that if colchicine solutions are injected into rabbits, the clotting time of the blood of these animals is markedly increased (LOICQ 1937). In plant tissues, the antimitotic action of colchicine seems also to be associated with a decreased viscosity of the cytoplasm; this can be concluded from the work of NORTHEN (1950) on onion root tips. However, in studies on the effect of very dilute solutions of colchicine on the protoplasmic viscosity of *Elodea* cells, STÅLFELT (1947; see also STÅLFELT 1956) found that very dilute solutions of colchicine caused an increase in the viscosity after the *Elodea* leaves had been immersed in the solutions for 18—65 hours.

In the chemical treatment of cancer, one of the most widely used substances is nitrogen mustard; this is methyl-bis (beta-chloroethyl) amine. Ordinarily the hydrochloride of nitrogen mustard is used. According to

Heilbrunn and Wilson (1957), a 0.1% solution of nitrogen mustard hydrochloride prevents the division of *Chaetopterus* eggs and it does this by keeping the protoplasm fluid and thus preventing the mitotic gelation. As a therapeutic agent, the oxide derivative of nitrogen mustard is probably superior to ordinary nitrogen mustard. This compound is manufactured by the Yoshitomi Pharmaceutical Laboratories in Osaka, Japan, and goes under the name of Nitromin. When fertilized *Chaetopterus* eggs were exposed to 0.2—0.5% solutions of Nitromin hydrochloride, the normal viscosity increase which precedes the appearance of the mitotic spindle did not occur; in other words, the protoplasm remained fluid and cell division was inhibited.

Another substance sometimes used in the treatment of cancer is 6-mercaptopurine. This substance is very sparingly soluble in sea water, but enough of it can be gotten into solution so as to prevent cell division in some, though not all, of the fertilized *Chaetopterus* eggs exposed to it. In these eggs also, the mitotic gelation tends to be inhibited.

Although the chemists and pathologists who have desribed many cases of the Haddow paradox have scarcely been able even to suggest a possible interpretation of it, an explanation is readily available on the basis of information that can be obtained from studies of protoplasmic viscosity. Nitromin, which in dilute solutions keeps the protoplasm fluid, in somewhat higher concentrations (0.5—1%) causes a sharp increase in the viscosity of the protoplasm of the unfertilized *Chaetopterus* egg. This viscosity increase is associated with a clotting reaction, that is to say the typical vacuolization reaction so often produced in marine eggs by agents which induce cell division. However in the few experiments tried by Heilbrunn and Wilson, no cleavage was induced by the solutions of Nitromin studied. This may have been due to the toxic effects of the substance. But in the case of urethane, which is well known to be both a carcinogenic agent and a therapeutic agent for cancer, it can readily be shown that whereas dilute solutions inhibit mitosis, more concentrated solutions are very effective in initiating mitosis. And, as was to expected, the dilute solutions keep the protoplasm fluid, whereas the more concentrated solutions cause a gelation of the protoplasm. Indeed, urethane, which is an anesthetic, acts like fat solvents in being able either to liquefy protoplasm or make it gel. And, as in the case of the fat solvents, this ability of urethane either to keep the protoplasm fluid or to gel it is almost certainly due to the fact that urethane can free calcium from combination with lipoprotein. This would tend to make calcium enter the interior of the cell from the cortex and also would tend to prevent any clotting action it might exert there. In other words, the same substance could have two opposite types of action.

There are liquefying substances which presumably do not act as fat solvents (and urethane) do, and do not release calcium from the cell cortex. The liquefying action of heparin has already been discussed. In this connection it is interesting to note that heparin has been found to have some value in improving the survival time of cancerous animals (Zakrzewski 1932. Balazs and Holmgren 1949). Much more effective in causing the regression of tumors (in mice) is the bacterial polysaccharide obtained from *Serratia marcescens*. This polysaccharide has been studied extensively at the National Cancer Institute in Bethesda, Maryland, by SHEAR and his group (see, for example SHEAR and PERRAULT 1943, and literature cited by them). SHEAR's polysaccharide prevents clotting of human blood although as a blood anticoagulant it is several hundred times weaker than heparin (MOST 1951). However it is much more powerful than heparin in preventing the mitotic gelation, and it also has a more powerful antimitotic action (HEILBRUNN and WILSON 1950 a and b).

Perhaps it is true that all types of living material contain substances which act like heparin in preventing protoplasmic clotting. Certainly heparin-like substances can be found in bacteria, protozoa, and in various types of tissue cells in both invertebrate and vertebrate animals. Some of these substances might be useful in the treatment of cancer. For some years now, HEILBRUNN and his associates have been searching for heparin-like antimitotic substances that might be useful in cancer therapy. Ovaries both of invertebrate and vertebrate animals are especially rich in such substances and they have been found in the ovaries of starfish, clams, lobsters, fishes, salamanders, frogs, chickens, rabbits, dogs, sheep, pigs and cows (HEILBRUNN, WILSON and HARDING 1951; HEILBRUNN, CHAET, DUNN and WILSON 1954; HEILBRUNN and WILSON 1956). These substances keep protoplasm fluid and prevent the mitotic gelation. Extracts from starfish ovaries are especially potent for marine eggs. On the basis of this information, it has been possible to obtain from cow ovaries extracts which markedly increase the survival time of mice implanted with a very lethal ascites tumor. Indeed in a long series of experiments on some thousands of cancerous mice, it was found that approximately 20% of the treated mice survived indefinitely, whereas in the untreated controls less than 1% survived and these were undoubtedly mice that were not properly inoculated (HEILBRUNN, WILSON, TOSTESON, DAVIDSON and RUTMAN 1957). More recent experiments with purified extracts give higher percentages of survival.

Thus on the basis of our knowledge of protoplasmic viscosity and the factors influencing it, it is possible to interpret the action of various agents which have been used in the attempt to cure cancer, and it may also be possible to develop new curative agents.

Various Drugs

The subject of pharmacology has had a long history, going back to the very beginnings of human knowledge. For even primitive man must have been interested in knowing to what extent it was possible to cure himself of disease, or to influence his various bodily functions by eating or drinking or exposing himself in one way or another to the various substances or mixtures of substances he was able to find in his environment.

For many long years pharmacology was essentially a practical discipline taught to medical students so that they might be better able intelligently to practice their art. With the development of pharmacology as a science, its first task, according to CLARK (1937) "was to sift the traditional beliefs of therapeutic practice and to discover how much of this was based on fact and how much was merely a legacy from mediaeval superstition." Modern pharmacologists are becoming increasingly eager to discover why it is that a drug acts, and they are becoming more and more cognizant of the fact that such knowledge to a large extent depends on an understanding of what a drug does to the living protoplasm.

In their famous book on "The Pharmacological Basis of Therapeutics," GOODMAN and GILMAN define a *drug* as "any chemical agent which affects living protoplasm," and in various parts of their huge volume they consider the possible effect or effects that individual drugs might have on cells. Unfortunately, however, burdened as they were with the enormous literature of their subject and concerned largely with practical matters, they were scarcely in a position to evaluate properly the work that has been done on the effects of drugs on protoplasm.

There are various ways in which drug action can be interpreted in terms of the physiology of the cell. For one thing, it is obvious that if a substance can not enter any part of a cell it could not very well have any (direct) effect on the protoplasm in the cell interior. That is of course why most cells can withstand a relatively high concentration of strong acids and bases, for the plasma membranes of most cells are rather impermeable to ions, and it is only when these membranes are destroyed that the strong acids and bases have much of an effect. In general the sensitivity of a cell to any drug must depend, to some extent at least, on the ability of the drug to pass through the plasma membrane. Although all cells, both plant and animal, show a somewhat similar permeability to various substances, it is also true that each individual type of cell differs from every other type of cell in the resistance it offers to the entrance of various chemical compounds. This variability has been long known and much studied. Some facts concerning it have been summarized by DANIELLI (1950). Perhaps the selective action of drugs on tissues will ultimately be explained, at least in part, by the permeability characteristics of the cells in the different tissues. It is also possible to interpret the action of drugs in terms of the effect they may have on the permeability of the plasma membrane. But in so far as our present knowledge goes, the attempt to explain drug action in terms of permeability relations has not been overly successful. Thus the old idea of explaining the action of anesthetics by some postulated effect on the permeability of the plasma membrane has had to be abandoned (for discussion see HEILBRUNN 1951, 1952).

Nor has the attempt to explain drug action in terms of effects on respiration or oxidizing enzymes, an attempt which has inspired a vast number of research efforts, been rewarded with much success. This is rather clearly demonstrated in the excellent review by HUNTER and LOWRY (1956). Often enough a drug may increase or decrease the rate of respiration, but as HUNTER and LOWRY point out, this is more likely to be the result rather than the cause of the drug action.

In view of the fact that no great measure of success has attended the efforts of those who have sought to interpret the action of drugs on cells in terms either of permeability effects or in terms of respiration and oxidizing enzymes, it is perhaps surprising that more work has not been done on the effects of drugs on the protoplasmic colloid. Certainly, as has been abundantly shown, changes in protoplasmic viscosity can serve as a very sensitive indicator of what is happening to the protoplasm when it is exposed to various physical and chemical influences.

In other sections of this article, attention has been directed to the manner in which various substances change the viscosity of the protoplasm. All these substances, in the GOODMAN and GILMAN definition, are drugs. But there are a vast number of other substances concerning which there is no information, or at best only a suggestion as to how they many affect the protoplasm. There is strong indication that many of the substances which have a very pronounced effect on vital processes do markedly alter the viscosity of the protoplasm. For substances like digitalis are relatively ineffective in the absence of calcium (see for example GOLD and KWIT 1937), and this is true also of adrenaline (LIBBRECHT 1920). Indeed adrenaline is known to release calcium from binding with protein (LAWACZEK 1928, HERMANN 1932). Cocaine likewise releases calcium from binding (BERWICK 1951). Moreover, acetylcholine seems to be ineffective in the absence of calcium (PUNT 1942; McNAMARA, KROP, and McKAY 1948), and this is also true of histamine (GILLESPIE and THORNTON 1932). These drugs may well produce their action by releasing calcium from the cortex into the cell interior. In studies on the oyster heart, OTIS (1942) concluded that the effects of adrenaline and of digitalin were due to such calcium release.

Obviously if these drugs are able to release calcium from the cell cortex, they must produce viscosity changes in protoplasm. But there is almost no direct evidence to indicate that this is true. What there is is certainly suggestive. BAUER (1926) exposed paramecia, amebae and frog leukocytes to very dilute solutions of histamine and adrenaline and then tried to see if these solutions had any effect on the protoplasmic colloid. His methods were rather crude. The cells were placed on a slide under a coverslip; he then broke them by pressing on the coverslip and watched the manner in which the protoplasm emerged. One part in 200,000 of histamine seemed to liquefy the protoplasm of the paramecia, and even one part in a million seemed in time to have some effect. Observation of Brownian movement also indicated liquefaction. On the other hand, adrenaline apparently had a gelating effect on the protoplasm. In this case also BAUER judged the viscosity by breaking the cells by pressing on the coverslip. Certainly these observations are not very impressive, and they perhaps relate more to the surface precipitation reaction than to the viscosity. Fortunately there is some supporting evidence for the gelating action of adrenaline. WENSE (1934) fed paramecia starch and then measured the viscosity of the protoplasm by noting how long it took for the starch-containing vacuoles to move under the influence of centrifugal force. One part in 100,000 of adrenaline caused a definite and easily observable increase in viscosity, and even one part in 150,000 seemed to have a similar effect.

It is strange indeed that more work has not been done on the effects of various hormones on the protoplasm of living cells. The efforts of biochemists to find out why and how these hormones act have not met with too much success. One of the most puzzling enigmas of endocrinological science is the problem of the basic reason or reasons for the action of insulin. A tremendous amount of effort has been expended in the attempt to solve this riddle, and some of the best minds in the field of biochemistry have struggled with the problem. The answer may not be found in a study of the action of insulin on isolated constituents of the cell, and indeed Levine, one of the most distinguished investigators of insulin action has been led to conclude that "It would be expected therefore that a certain degree of morphological intactness is necessary to demonstrate hormonal effects and actions. Otherwise we would be pulling the trigger of an unassembled gun" (Levine and Goldstein 1955). Levine and Goldstein offer evidence to show that insulin acts by virtue of an effect it has on the active transport of sugar into the cell, but this evidence is somewhat indirect and is not as convincing as it might be.

Some years ago, Heilbrunn and Wilson attempted to determine the effect of insulin on the protoplasmic viscosity of the eggs of the worm *Chaetopterus*. They found a viscosity increase, but the preparations of insulin that they used contained sufficient zinc to make the result obtained questionable. More recently through the kindness of the Eli Lilly Company it has been possible to obtain a preparation of insulin containing extremely small amounts of zinc. With this preparation Heilbrunn, Ashton and Feldherr are making studies of the action of insulin on the protoplasm of the giant ameba *Chaos chaos*. In this ameba gel-sol and sol-gel transformations go on rapidly. The gelation that occurs undoubtedly involves the clotting reaction so common to protoplasm, a reaction similar to the clotting of blood (see section on Clotting and Anticlotting Substances). As previously noted, heparin can readily be shown to prevent this reaction in *Chaos chaos*. The interesting fact is that the anticlotting effect of heparin is counteracted by dilute solutions of insulin. Moreover the strong metachromatic color normally given by the ameba protoplasm when exposed to toluidine blue, a color reaction presumably due to heparin or heparin-like substances, does not appear in amebae previously treated with dilute solutions of insulin. The significance of these preliminary experiments is great. For if as we have been led to believe (Heilbrunn 1956 a, Wilson and Heilbrunn 1957), the protoplasmic clotting reaction sets off a train of oxidative processes, in the colloidal reaction of ameba protoplasm to insulin we may have a clue as to the essential reason why insulin acts as it does.

In plants there are substances that act in somewhat the same way as do the hormones of animals; that is to say there are substances produced by cells in one part of the organism which act on cells in another part of the organism. These are the auxins. There are many substances which are known to have auxin activity. Northen (1942) smeared the petioles and stems of bean plants with lanolin containing indole-3-acetic acid, indole-3 n-propionic acid, and alpha-naphthalene acetic acid. All of these substances are auxins. NORTHEN studied the effect on protoplasmic viscosity by centrifuging the petioles and stems and then sectioning them so that he could observe the movement of starch grains. He compared the viscosity of the protoplasm of the cells on the side on which the auxin-containing grease had been smeared with the viscosity of the protoplasm of the cells on the control side. In general, the auxins caused a lowering of viscosity. NORTHEN believes the effect to be due to a splitting of protein molecules.

STÅLFELT (1947) exposed *Elodea* leaves to various concentrations of indole-acetic acid (heteroauxin). Even in extremely low concentration, this sub-

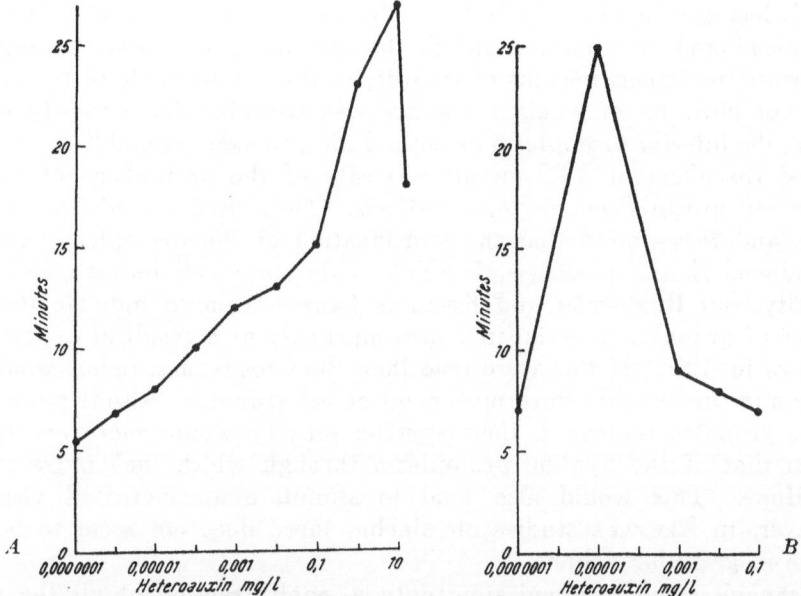

Fig. 21. The effect of various concentrations of heteroauxin on the viscosity of the protoplasm of *Elodea* leaves. The ordinates show the number of minutes of centrifugation time required to displace the chloroplasts in all the cells on the upper side of the leaf. Abscissae show mg. per liter of heteroauxin. *A* shows the viscosity values 24 hours after exposure; *B* after 38 hours.
(STÅLFELT.)

stance caused an increase in the viscosity of the protoplasm as determined by centrifuge tests. The increase was considerable, but in the relatively higher concentrations, after some hours, the viscosity instead of continuing to increase progressively, began to decrease again. Two of STÅLFELT's graphs are shown in Fig. 21. One remarkable fact is that a concentration as low as one part per million could have a measurable effect on the viscosity of the protoplasm.

In view of the fact that different concentrations of auxins can either increase the rate of growth or depress the rate, it is not surprising that these substances may either increase or decrease the viscosity of the proto-plasm. Both NORTHEN and STÅLFELT call attention to the fact that changes in viscosity occur during cell division, and STÅLFELT concludes his paper by supporting the idea that the primary factor in the effect of heteroauxin is its action on protoplasmic viscosity.

For the modern physiologist and biochemist, one of the most interesting of all substances is adenosine triphosphate or ATP. Ever since Szent-Györgyi (1949) was able to show that rabbit muscle fibers dehydrated in glycerol, and long dead, would shorten when exposed to ATP, students of the physiology of muscle have been greatly intrigued by this substance, even though it has but little effect when injected into a *living* muscle fiber (Wiercinski, Cookson and Lajtha, cited by Heilbrunn 1952; Falk and Gerard 1954).

Kriszat (1950) studied the effect of ATP on the viscosity of the protoplasm of the giant ameba *Chaos chaos*. He found that ATP causes an initial decrease in viscosity followed by an increase. However it should be remembered, as pointed out in the section on protozoan protoplasm, that centrifuge measurements of viscosity in this ameba are hard to evaluate. It is not clear as to whether Kriszat was studying the viscosity of the cortex, the interior protoplasm or both. Later, Runnström and Kriszat (1950) studied the effect of ATP on the viscosity of the protoplasm of the egg of the sea urchin *Psammechinus miliaris*. They used a centrifuge microscope, and they found that the stratification of the protoplasm was less pronounced in the presence of ATP. This may well indicate increased viscosity, but Runnström and Kriszat's figures seem to indicate that the volume of granular material increases markedly as a result of exposure of the eggs to ATP. If this were true then the Cunningham factor would increase and this would retard movement of the granules. Also if the volume of the granules increased, their specific gravity would more nearly approach that of the hyaline protoplasm through which they move on the centrifuge. This would also tend to stimulate an increased viscosity. However, in Kriszat's studies on ameba, there does not seem to be any change in granular volume.

Although at the present time only a small fraction of all the drugs used by the clinician and studied by the pharmacologist have been examined as to their effects on the protoplasmic colloid, it is to be hoped that in the not too distant future there will be an awakening of interest in this field. We need much more information than we now have as to what drugs actually do to the living cells they act upon.

Injury and Injury Substances

Just as the science of pharmacology could profit from a better understanding of the action of drugs on protoplasm, so too the science of pathology is in need of additional information as to what happens to protoplasm as a result of injury. Various pathological conditions are due to changes in the protoplasmic colloid, and many modern pathologists are beginning to realize that if one is properly to understand degenerative changes in protoplasm, one can not depend for information solely on studies made on cells whose protoplasm has been thoroughly coagulated and then cut into thin slices.

In earlier discussion, mention has already been made of the effects of cold and heat injury, and of the effects on the protoplasmic colloid of injurious doses of radiation. The results of exposure of cells to excessive doses of toxic substances have also been considered. In this section we shall be concerned chiefly with the nature and mode of action of substances produced by injured cells, substances which have an effect on cells at some distance from the site of the original injury. This is an extremely important topic both for the pathologist and the clinician, for in man and other highly organized animals, whenever some cells are injured in one way or another, they produce substances which have an effect on other cells and tissues at a distance, so that severe consequences can result and death may follow. Thus if a man receives a sharp blow on the head, if his muscles are crushed, or if any large part of his body is burned, he will go into a state of shock from which he may not recover. These are well known facts, and the literature on traumatic shock, burn shock and other types of shock, is enormous. At the present time, it is generally conceded that an important factor in the causation of shock is a toxic substance or substances that are given off; the technical name for them is "toxic factor." Of course in a higher animal there are reactions and counterreactions going on in the tissues and in the blood stream, so that it is difficult to understand the course of events that follow the entrance of toxic factor into the blood of an injured animal.

In plants the phenomena are simpler. Bünning (1926) experimented on seedlings of rye and radish and on shoots of *Tradescantia viridis*. He cut the tissues of these plants and then with the centrifuge method studied the viscosity of the protoplasm of the cells in the vicinity of the wound. In these cells there was a progressive and pronounced increase in protoplasmic viscosity; this increase was greater in the cells nearer the wound. In the *Tradescantia* plant the maximum viscosity increase (of about 250%) occurred in about a minute; in the rye seedling the increase occurred somewhat more slowly—in about 3½ minutes the viscosity had risen to at least 5 or 6 times its original value.

When a *Spirogyra* filament is cut or injured locally with a hot needle; and then centrifuged, the chloroplasts in the cells near the wound are more readily displaced than in the cells remote from the wound (Northen and Northen 1938). This is clearly an indication of the release of injury substances from the wounded cells, as was pointed out by Umrath (1942), who repeated and confirmed Northen and Northen's experiments. The more rapid displacement of chloroplasts in *Spirogyra* may be due to a decrease in viscosity or liquefaction either of the cortex or of the interior protoplasm or both. If we assume that in these injury experiments it is the cortex that is liquefied, the experiments of Northen and Northen and of Umrath are not in conflict with those of Bünning. As is well known, injury to a plant causes an increase in cell division in nearby cells. The viscosity changes that occur in the cells studied by Bünning would be of the sort that would initiate mitosis. And if in the experiments on *Spirogyra* the cortex was primarily involved in the viscosity decrease, then this would also be a condition favorable to the onset of cell division (compare Wilson and Heilbrunn 1952).

When animal cells are injured, they also are capable of giving off injury substances. Heilbrunn, Harris, LeFevre, Wilson and Woodward (1946) studied extracts from a wide variety of heat-killed organisms and tissues (sea anemones, worms, clams, squids, lobster muscle, fish muscle and frog muscle). All of these extracts contained substances which could initiate a clotting reaction, that is to say a surface precipitation reaction, in the protoplasm of sea urchin eggs broken in the absence of calcium. Hence they were thought to contain thrombin-like substances. In many instances if the intact eggs were exposed to these extracts, the protoplasm showed a large increase in viscosity. This increase in viscosity seemed to proceed from the periphery of the cell inward as the clotting agent slowly diffused into the interior. In other cases the clotting substance did not appear to be able to penetrate the cell membrane, and it had no effect on the protoplasm of the intact egg cells.

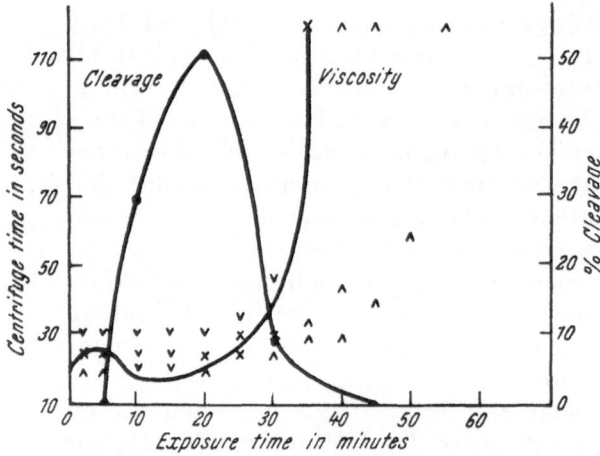

Fig. 22. Protoplasmic viscosity and percent cleavage of *Arbacia* eggs after exposure for various lengths of time to an extract of frog muscle made after the muscle had been heated to 50° C. Viscosity determinations were made immediately after the removal of the eggs to sea water. *V* indicates that the eggs were stratified; *Λ* that the eggs were not stratified; and *X* that the eggs were partially stratified.
(D. Harding.)

In experiments on sea urchin (*Arbacia*) eggs, placed in extracts of heat-killed frog or clam muscle, Drusilla Harding (1951) found that the viscosity of the protoplasm increased markedly as a result of exposure to the injury substances contained in the extract. This viscosity increase proceeded slowly, either due to the slow rate of diffusion of the potent substance of the extract or to the slowness of the reaction. If eggs were removed from these solutions before the increase in viscosity had gone too far (or had even begun), a rather high percentage of the eggs cleaved. This is shown in Fig. 22, taken from Harding's paper.

In all probability, the injury substances given off by various types of living material differ, and it is rather doubtful if an injured plant releases the same substances as an injured muscle. But there is an underlying similarity in the response of various cells and tissues to injury. This is evidenced first by the fact that the injured regions so commonly have a negative electric potential as compared with the uninjured regions. And it is shown even more clearly by the fact that injured cells so frequently show the same type of colloidal reaction in their protoplasm. When cells are severely injured, whether by mechanical trauma, by excessive heat, or cold, or radiation, their protoplasm commonly undergoes a clotting reaction,

a reaction which is evidenced morphologically by the appearance through
the protoplasm of vacuoles or fibrils. This clotting reaction will be discused
more fully in the next section. Suffice to say here that it is a reaction
similar to the clotting of blood. When blood clots, as one consequence of
the reaction, a substance is formed which can cause unclotted blood to clot
even in the absence of calcium. This is the well known enzyme thrombin,
which is both a clotting enzyme and a proteolytic enzyme. Hence when
the protoplasm of a cell is injured and clots, we might expect that it would
produce substances which would have a clotting effect on the protoplasm
of other cells. This point of view was first clearly stated by HEILBRUNN
(1937, on page 458). Since then, there is increasing and many-sided

evidence that when the cells of
vertebrate animals are injured,
there is a great increase in protease activity, and presumably
there is an increase in the proteases which find their way into
the blood. It is scarcely necessary
to review the literature in this field.
A recent and interesting paper is
that of UNGAR and DAMGAARD (1954)
and this may be consulted for references to earlier papers. UNGAR
and DAMGAARD found that if gui-
nea pig lung tissue was either
heated or chilled, there was a great
increase in protease activity. This
could be demonstrated by making
extracts of the heated and cooled
tissue and comparing the proteo-

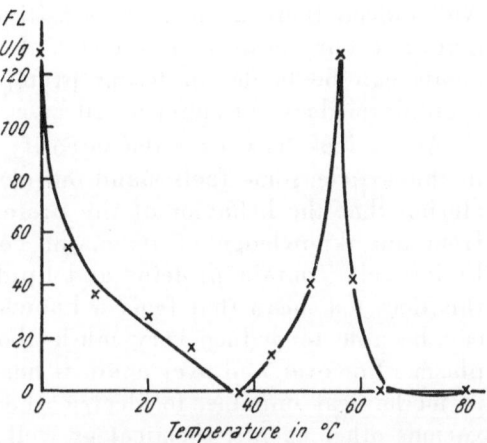

Fig. 23. Protease activity of extracts from heated and
cooled lung tissue. Abscissae: temperatures to which lung
slices were exposed before extraction. Ordinates: fibrino-
lytic units per gram of lung.
(UNGAR and DAMGAARD.)

lytic activity of these extracts with the proteolytic activity of extracts made
from the tissue at body temperature. Fig. 23 shows the results obtained.

Proteases cause clotting of blood and they can also cause a clotting in
protoplasm (see discussion in section on Clotting and Anticlotting Substances). Moreover, intravenous injection of proteases into the blood stream
of an animal can cause shock. Various papers have been published to
show this (for literature references see WEIMAR 1955 a and b). An increase
in protease activity in the blood stream seems to initiate a homeostatic
reaction to counteract the clotting activity of the proteases. This is apparently the reason why heparin or some heparin-like substance comes to be
present in the blood of shocked animals. Apparently much of the distress
in shock is due to the presence of such a substance (compare RIESER 1955 a).

Clearly the cell physiologist in his studies of protoplasmic viscosity
and of the colloidal state of injured cells and of cells exposed to the action
of injury substances has contributed something to the fundamental back-
ground of the science of pathology, and in the future it is hoped that he
will contribute a great deal more.

The Significance of the Viscosity Data

What can be learned from the viscosity data that have been presented? Certainly if students of protoplasmic viscosity were content merely to discover whether this or that condition or this or that agent caused an increase or a decrease in the viscosity, not very much would be gained from the enterprise. Surely the main point in making viscosity measurements is to find out as much as possible about the protoplasmic colloid, what it is like in the resting cell, how it changes during activity, what factors are involved in its deterioration with age, and how it is affected by various chemical and physical agents. At the present time there is no better way of studying the protoplasmic colloid than to measure its viscosity. And indeed there is no more sensitive indicator of the physical changes that are occurring in the living cell. Fortunately, proper viscosity measurements can be made on living protoplasm; and generally speaking, the protoplasm does not suffer at all as a result of the measurement.

As we look back over the viscosity studies that have been summarized in this article, some facts stand out clearly. First there is the certain conclusion that the behavior of the protoplasmic colloid can not be deduced from any knowledge of its major constituents as we now know them. Living cells contain proteins and lipids and carbohydrates and salts, but this does not mean that from a knowledge of the behavior of proteins we can be able to deduce very much about the colloidal behavior of protoplasm. For over and over again it has been shown that protoplasm reacts to mechanical impacts, to electric shocks, to ultraviolet radiation, and to various other agents, chemical as well as physical, in a far different way from the way in which solutions of proteins or of proteins mixed with lipids would react. Chemists prefer to work with pure substances, but the behavior of a colloid is often largely dependent on the presence of small amounts of substances which the conscientious chemist would try to eliminate as impurities. Hence to extract a purified protein from a cell and to discover its colloidal behavior may tell very little about the colloidal behavior of the same protein when it is in the cell with all the numerous other constituents of the protoplasm. Moreover the cell has organization, and it is only when this organization is intact and functional that life is present. And the student of protoplasmic viscosity and of the protoplasmic colloid is never much interested in the behavior of what may once have been a living cell.

In the study of the viscosity of the protoplasm of a living cell, there are certain difficulties. Fortunately, injury to the protoplasm is ordinarily not one of them. In some instances the force of gravity is sufficient to cause inclusions of the cytoplasm to move, and from this rate of movement the viscosity can be calculated. When it is necessary to use centrifugal force to move granules through the cytoplasm, the amount of centrifugal force required is usually far less than that which will cause injury. Indeed many types of cells can withstand enormous centrifugal forces. Moreover, in some cases at least, the viscosity values obtained are the same when determinations are made over a wide range of centrifugal forces. This could scarcely be true if the centrifugal force caused serious injury or had some effect in altering the viscosity. However, it should perhaps be recognized that almost any method of measuring viscosity even of lifeless fluids might conceivably alter the viscosity while the measurement was in progress. Thus the passage of a liquid through a tube might have an effect on the liquid whose viscosity was being measured. This was clearly recognized by the physical chemist DRUCKER (1917). However, no one doubts but that the values for viscosity that have been published by physicists and physical chemists are essentially accurate. And in the measurement of protoplasmic viscosity by the Brownian movement method, it would scarcely be possible for the movement of the particles to have any but an infinitesimal effect on the viscosity. The fact that measurements made by a study of Brownian movement are in essential agreement with those made by centrifuging cells is encouraging evidence that both methods, if properly used, can give results of the correct order of magnitude.

In many diverse kinds of living cells, there is a stiff outer cortex surrounding an interior protoplasm which is fluid and which may be highly fluid. In ameba, it possible by using different species to make measurements of the viscosity of the outer cortical protoplasm and of the inner more fluid protoplasm. The two are very different. Moreover they respond differently to various treatments. Thus a small amount of calcium will markedly stiffen the protoplasm of the cortex and will have quite an opposite effect on the protoplasm of the interior. Also stimulating agents such as ultraviolet radiation or mechanical impacts may have opposite effects on the cortex and the interior. This is true not only of the protoplasm of the ameba but of the protoplasm of marine eggs also. In some types of cells it is not possible to decide whether the viscosity determinations are for the cortical protoplasm, the interior protoplasm, or both. So for example, if filaments of the alga *Spirogyra* are centrifuged, the ease with which the chloroplasts can be moved depends in part on how readily they can be detached from the more rigid cortex, but presumably it also depends on the viscosity of the inner protoplasm through which they are shifted to one end of the cell. Hence in considering the significance of viscosity measurements on this type of material, it is not possible to decide with any certainty whether changes observed are due to changes in cortex, interior or both. This is a serious limitation.

In some types of cells, paramecium for example, the protoplasm contains so very many granules that the whole mass constitutes a highly concentrated suspension. Colloid chemists have found in their studies of suspensions of this type that the viscosity of the suspension as a whole varies with the shearing force and is greater the greater the shearing force. In other words, the suspensions exhibit the phenomenon known as dilatancy. It is not surprising therefore that the protoplasm of paramecium is dilatant, and if a strong shearing force is applied to it, the viscosity is indeed very great. However the viscosity of the hyaline protoplasm in which the granules are suspended is undoubtedly very low, for such low viscosity is a characteristic of dilatant systems. It is probable that in all types of cells containing a high concentration of granules, the protoplasm as a whole is dilatant, but as yet we have no data for protoplasm other than that of paramecium.

Let us turn now to our major question. What sort of a colloid is protoplasm, and how can we possibly interpret its strange patterns of behavior? In the light of our present knowledge, the answer seems clear. In the cells that have been studied, the protoplasmic colloid is a delicately poised system subject to sudden sharp changes in viscosity, changes which are related in an intimate way with vital phenomena. If within a cell the concentration of free calcium ions rises even by a rather small amount, this can initiate a clotting reaction, a reaction which can occasion a sharp increase in protoplasmic viscosity, can even make the viscosity become infinite. The clotting reaction in protoplasm is essentially similar to the clotting reaction that occurs in vertebrate blood. This is not all surprising, for the substances involved in blood clotting are all substances that have been produced by living cells. And from various types of cells substances can be extracted which can either hasten or delay the clotting of vertebrate blood. The great pioneer in the study of blood clotting, ALEXANDER SCHMIDT, knew that from all types of cells, plant as well as animal, it was possible to obtain substances which would markedly hasten the clotting of horse blood (SCHMIDT 1892, 1895). And it is likewise possible to extract from various types of cells substances which act like heparin in slowing or preventing the clotting of vertebrate blood. A striking example of this is the fact that from the tissues of the clam, an animal whose blood does not clot, a heparin-like substance can be obtained, a substance which is a more powerful anticoagulant for sheep blood than ordinary heparin from mammalian tissues (THOMAS 1951, 1954; FROMMHAGEN, FAHRENBACH, BROCKMAN and STOKSTAD 1953).

The irritability of protoplasm, so evident in certain types of living material, can be interpreted on a colloid chemical basis. For there are many agents, the so-called stimulating agents, which are able to liberate calcium ions from the cortex of the cell. This release of calcium initiates a clotting reaction in the cell interior, a reaction commonly preceded by a slight decrease in viscosity as the first very small amounts of calcium enter. To appreciate how very sensitive the protoplasm of a cell can be to small amounts of calcium, it is only necessary to expose the interior protoplasm to a trace of calcium. This can be done either by breaking the cell so as to rupture its outer membrane, or by injecting calcium into the cell interior. The effect is especially striking in muscle protoplasm, for as this protoplasm clots there is a pronounced shortening (HEILBRUNN 1940, WOODWARD 1948, HEILBRUNN and WIERCINSKI 1947).

In blood the reactions involved in the clotting process are extremely complicated and there is no doubt but that they are extremely complicated in protoplasm also. Because our knowledge of protoplasmic clotting is so rudimentary as compared with the tremendous amount of information about the clotting of blood, it is scarcely possible to institute anything like a complete comparison between the two processes. One essential fact about blood clotting is the fact, universally recognized, that it occurs in two stages. The first of these requires calcium; the second does not. The first stage or series of reactions results in the activation of a clotting enzyme, thrombin, which is also a protease. In protoplasmic clotting also there appear to be two stages. For even without the addition of calcium ion it is possible to clot protoplasm by injecting minute traces of a protease like trypsin. Also clotted protoplasm can produce substances which can initiate a clotting process in other protoplasm in the absence of calcium (HEILBRUNN 1927).

Another similarity between the clotting of blood and protoplasm is the fact that in both types of colloidal reactions, the presence of the mucopolysaccharide heparin or some kindred substance can slow or inhibit the reaction. Heparin or heparin-like substances are present in many, perhaps in essentially all, types of cells (for evidence see HEILBRUNN 1956 a).

If cells generally contain precursors of clotting enzymes similar to thrombin and mucopolysaccharides similar to heparin, in the interplay of these two types of substances it is possible to conceive of a mechanism or mechanism which could serve to interpret much of the machinery of vital processes. In any cell the clotting substances may be devoid of activity until activated by calcium; and the inhibitors of the clotting reaction, substances like heparin, may be inactive because of combination with protein. A protease could begin to clot protoplasm but by its very proteolytic action it might free its own inhibitors. Such a cyclic series of changes is a necessary concept in the interpretation of any protoplasmic system capable of doing mechanical work, as in the contraction of a muscle or the division of a cell.

In the past, biochemists have been in the habit of explaining the mechanical energy of protoplasm in terms of its oxidative mechanisms, but they have never arrived at a satisfactory explanation of how an oxidative process can perform mechanical work, and one by one their attempts to harness oxidative chemical reactions to the mechanical energy of the living engine have failed. In such theorizing it should be remembered that cells which do no obvious mechanical work also have vigorous oxidative processes. These processes furnish the energy for chemical syntheses which enable cells to build up complex organic compounds. Granted the formation or the existence of these organic compounds, in the energy latent in their huge molecules they can furnish the motive power for the colloidal processes necessary for the contraction of a muscle or for other vital phenomena that require the expenditure of energy. Often, though not always, when living cells are aroused or activated, their oxidative processes increase. This can logically be interpreted, as WILSON and HEILBRUNN (1957) have done, by the fact that the clotting of protoplasm involves a change in which SH groups are oxidized to S-S groups. Such a change might well trigger a whole succession of oxidative reactions. But the primary impulse that arouses a cell is certainly not some effect on oxidative enzymes; at any rate no one even by the wildest sort of speculation has

7*

been able to suggest how the various different types of stimuli which exite protoplasm could possibly have a direct effect on the activity of oxidative enzymes.

When we come to know more about the viscosity of protoplasm, how it is affected by one drug or another, how it is influenced by untoward changes in the cellular environment, we shall certainly be able to arrive at a more complete and a more rational understanding of why drugs act as they do, and how cells and the tissues of which they are composed react to injury. To understand the living colloid we must make studies of it while it is alive and intact, and not a jumble of unassembled and damaged parts. Perhaps in the near future there will be somewhat more emphasis on the colloid chemistry and the dynamics of living protoplasm, for only by a study of the interplay of the various substances and constituents of the living cell can we come to understand the basic facts about vital processes.

Bibliography

Allen, J. G., M. Sanderson, M. Milham, A. Kirschon, and L. O. Jacobson, 1948: Hyperheparinemia (?). An anticoagulant in the blood of dogs with hemorrhagic tendency after total body exposure to roentgen rays. J. exper. Med. (Am.) 87, 71—86.

Alsup, F. W., 1942: The effects of light alone and photodynamic action on the relative viscosity of *Amoeba* protoplasm. Physiol. Zool. 15, 168—183.

Angerer, C. A., 1936: The effects of mechanical agitation on the relative viscosity of amoeba protoplasm. J. cellul. a. comp. Physiol. (Am.) 8, 329—345.

— 1937: The effect of salts of heavy metals on protoplasm I. The action of cupric chloride on the viscosity of sea urchin eggs. J. cellul. a. comp. Physiol. (Am.) 10, 183—197.

— 1939: The effect of electric current on the relative viscosity of sea-urchin egg protoplasm. Biol. Bull. (Am.) 77, 399—406.

— 1942: The action of cupric chloride on the protoplasmic viscosity of *Amoeba dubia*. Physiol. Zool. 15, 436—442.

— and K. M. Wilbur, 1943: The action of various types of electric fields on the relative viscosity of *Amoeba proteus*. Physiol. Zool. 16, 84—92.

Arnold, H. D., 1911: Limitations imposed by slip and inertia terms upon Stokes' law for the motion of spheres through liquids. Phil. Mag. (6) 22, 755—775.

Ashton, F. T., 1957: Magnetic studies on cells and protoplasm. Biol. Bull. (Am.) 113, 319.

Austin, J. H., F. W. Sunderman, and J. G. Camack, 1927: Studies in serum electrolytes. II. The electrolyte composition and the pH of serum of a poikilothermous animal at different temperatures. J. biol. Chem. (Am.) 72, 677—685.

Baas-Becking, L. G. M., H. v. d. Sande Bakhuyzen and H. Hotelling, 1928: The physical state of protoplasm. Verhandl. Koninkl. Akad. Wetenschap. Amsterdam, Afdeel. Natuurk. (Tweede Sectie). 25, 1—53.

Balazs, A., and H. Holmgren, 1949: Effect of sulfomucopolysaccharides on growth of tumor tissue. Proc. Soc. exper. Biol. a. Med. (Am.) 72, 142—145.

Barr, G., 1931: A monograph of viscometry. New York.

Barth, L. G., 1929: The effects of acids and alkalies on the viscosity of protoplasm. Protoplasma 7, 505—534.

Bauer, V., 1926: Über die Wirkung von Histamin und Adrenalin auf Protozoen. Verh. d. dtsch. Zool. Ges. (Zool. Anz. Supp. 2), 172—177.

Bayliss, W. M., 1920: The properties of colloidal systems IV. Reversible gelation in living protoplasm. Proc. roy. Soc. (Lond.) B 91, 196—201.

Beams, H. W., 1949: Some effects of centrifuging upon protoplasmic streaming in *Elodea*. Biol. Bull. (Am.) 96, 246—256.

— and T. C. Evans, 1940: Some effects of colchicine upon the first cleavage in *Arbacia punctulata*. Biol. Bull. (Am.) 79, 188—198.

Bersa, E., and F. Weber, 1922: Reversible Viskositätserhöhung des Cytoplasmas unter der Einwirkung des elektrischen Stromes. Ber. dtsch. bot. Ges. 40, 254—258.

BERSIN, T., 1935: Thiolverbindungen und Enzyme. Erg. Enzymforsch. 4, 68—101.
BERWICK, M. C., 1951: The effect of anesthetics on calcium release. J. cellul. a. comp. Physiol. (Am.) 38, 95—107.
BIEBL, R., 1942: Wirkung der UV-Strahlung auf *Allium*-Zellen. Protoplasma 36, 491—513.
— 1947: Strahlenwirkungen auf das pflanzliche Protoplasma. Mikroskopie 2, 328—335.
BINGHAM, E. C., 1922: Fluidity and Plasticity. New York.
BODENSTEIN, D., 1947: The effects of nitrogen mustard on embryonic amphibian development. I. Ectodermal effects. J. exper. Zool. 104, 311—341.
BRECKHEIMER-BEYRICH, H., 1949: Über die Wirkung zentrifugaler Kräfte auf das Protoplasma von *Nitella flexilis*. Ber. dtsch. bot. Ges. 62, 55—60.
BRINLEY, F. J., 1928: The effect of chemicals on viscosity of protoplasm of amoeba as indicated by Brownian movement. Protoplasma 4, 177—191.
BROWN, R. H. J., 1940: The protoplasmic viscosity of *Paramecium*. J. exper. Biol. 17, 317—324.
BRÜCKE, E., 1862: Über die sogenannte Molekularbewegung in thierischen Zellen, insonderheit in den Speichelkörperchen. S.ber. Akad. Wiss. Wien, math.-naturw. Kl. 45 (*part* 2), 629—642.
BÜNNING, E., 1926: Untersuchungen über die Koagulation des Protoplasmas bei Wundreizen. Bot. Archiv 14, 138—164.
CHAMBERS, R., and C. KAO, 1951: The physical state of the axoplasm in situ in the nerve of the squid mantle. Biol. Bull. (Am.) 101, 206.
CHIFFLOT, J., and C. GAUTIER, 1905: Sur le mouvement intraprotoplasmique à forme Brownienne des granulations cytoplasmiques. J. Bot. 19, 40—44.
CHOLODNYJ, N., 1923: Zur Frage über die Beeinflussung des Protoplasmas durch mono- und bivalente Metallionen. Beih. bot. Cbl. 39, part 1, 231—238.
CLARK, A. J., 1937: General Pharmacology. Printed as a supplement to Heffters Handb. d. exp. Pharmakologie. Berlin.
COSTELLO, D. P., 1934: The effects of temperature on the viscosity of *Arbacia* egg protoplasm. J. cellul. a. comp. Physiol. (Am.) 4, 421—433.
— 1939: The volumes occupied by the formed cytoplasmic components in marine eggs. Physiol. Zool. 12, 13—20.
CRICK, F. H. C., and A. F. W. HUGHES, 1950: The physical properties of cytoplasm. A study by means of the magnetic particle method. Exper. Cell Res. 1, 37—80.
CUNNINGHAM, E., 1910: On the velocity of steady fall of spherical particles through fluid medium. Proc. roy. Soc. (Lond.) A 83, 357—365.
DANIELLI, J. F., 1950: Cell Physiology and Pharmacology. New York.
DAUGHERTY, K., 1937: The action of anesthetics on amoeba protoplasm. Physiol. Zool. 10, 473—483.
DRUCKER, C., 1917: Untersuchungen über Fluidität. Z. physikal. Chem. 92, 287—319.
DULBECCO, R., 1955: Photoreactivation. In Hollaender's Radiation Biology, vol. 2, New York. pp. 455—486.
DURYEE, W. R., 1939: Does the action of X-rays on the nucleus depend upon the cytoplasm? Biol. Bull. (Am.) 77, 326.
— 1949: The nature of radiation injury to amphibian cell nuclei. J. nat. Cancer Inst. 10, 735—795.
EIBL, K., 1939: Das Verhalten der *Spirogyra*-Chloroplasten bei Zentrifugierung. Protoplasma 33, 73—102.
EIGSTI, O. J., and P. DUSTIN, Jr., 1955: Colchicine—in agriculture, medicine, biology and chemistry. Ames, Iowa.
EINSTEIN, A., 1905: Über die von der molekularkinetischen Theorie der Wärme geforderte Bewegung von in ruhenden Flüssigkeiten suspendierten Teilchen. Ann. Physik (4) 17, 549—560.
— 1906 a: Eine neue Bestimmung der Moleküldimensionen. Ann. Physik (4) 19, 289—309.
— 1906 b: Zur Theorie der Brownschen Bewegung. Ann. Physik (4) 19, 371—381.
— 1911: Berichtigung zu meiner Arbeit: „Eine neue Bestimmung der Molekül-dimensionen." Ann. Physik (4) 34, 591—592.
— 1920: Bemerkung zu der Abhandlung von W. R. HESS „Beitrag zur Theorie der Viskosität heterogener Systeme." Kolloid-Z. 27, 137.
ERRERA, M., 1957: Effets biologiques des radiations. Aspects biochimiques. In Protoplasmatologia. Vienna.

Euler, H. von, and L. Hahn, 1946: Influence of Roentgen rays on isolated cell nuclei. Acta Radiol. **27**, 269—280.

Ewart, Alfred J., 1903: On the Physics and Physiology of Protoplasmic Streaming in Plants. London.

Falk, G., and R. W. Gerard, 1954: Effect of micro-injected salts and ATP on the membrane potential and mechanical response of muscle. J. cellul. a. comp. Physiol. (Am.) **43**, 393—403.

Fauré-Fremiet, E., 1913: Le cycle germinatif chez l'*Ascaris megalocephala*. Arch. Anat. microsc. **15**, 435—757.

Feichtinger, N., 1933: Viskositätsänderung des Protoplasmas als Folge radio-aktiver Bestrahlung. Naturw. **21**, 569—575, 589—591.

Fetter, D., 1926: Determination of the protoplasmic viscosity of *Paramecium* by the centrifuge method. J. exper. Zool. **44**, 279—283.

Fischer, H., 1950: Über protoplasmatische Veränderungen beim Altern von Pflanzenzellen. Protoplasma **39**, 661—676.

Flaig, J. V., 1947: Viscosity changes in axoplasm under stimulation. J. Neurophysiol. **10**, 211—221.

Forbes, A., and C. Thacher, 1925: Changes in the protoplasm of *Nereis* eggs induced by β-radiation. Amer. J. Physiol. **74**, 567—578.

Frederikse, A. M., 1932: Spontane Wiederherstellung der ursprünglichen Protoplasmaviskosität nach Erhöhung derselben unter Einfluß von Essigsäure. Protoplasma **15**, 603—611.

— 1933: Viskositätsänderungen des Protoplasmas während der Narkose. Protoplasma **18**, 194—207.

Friedenwald, J. S., W. Buschke, and R. O. Scholz, 1948: Effects of mustard and nitrogen mustard on mitotic and wound healing activities of the corneal epithelium. Bull. John Hopkins Hosp. **82**, 148—160.

Frommhagen, L. H., M. J. Fahrenbach, J. A. Brockman, and E. L. R. Stokstad, 1953: Heparin-like anticoagulants from mollusca. Proc. Soc. exper. Biol. a. Med. (Am.) **82**, 280—283.

Fürth, R., 1930: Über die Messung der Viskosität sehr kleiner Flüssigkeitsmengen mit Hilfe der Brownschen Bewegung. Z. Physik **60**, 313—316.

Gersch, M., and W. Möbius, 1951: Beitrag zur biologischen Wirkung kurzwelliger Strahlen (Untersuchungen an Wirbellosen). Strahlenther. **85**, 565—580.

Gibbs, R. D., 1926: Action of ultraviolet light on *Spirogyra*. Trans. roy. Soc. Canada, Sec. V, **20**, 419—426.

Giese, A. C., 1946: Comparative sensitivity of sperm and eggs to ultraviolet radiations. Biol. Bull. (Am.) **91**, 81—87.

Gillespie, M., and J. W. Thornton, 1932: The effect of calcium on the response of isolated bronchi to histamine. J. Pharmacol. (Am.) **45**, 419—426.

Gold, H., and N. Kwit, 1937: Digitalis and calcium synergism. Science **86**, 330—331.

Goldforb, A. J., 1935: Viscosity changes in ageing unfertilized eggs of *Arbacia punctulata*. Biol. Bull. (Am.) **68**, 191—206.

Goldstein, L., 1953: A study of the mechanism of activation and nuclear breakdown in the *Chaetopterus* egg. Biol. Bull. (Am.) **105**, 87—102.

Goodman, L. S., and A. Gilman, 1955: The Pharmacological Basis of Therapeutics. 2nd edition, New York.

Gray, J., 1927: The mechanism of cell-division. IV. The effect of gravity on the eggs of *Echinus*. Brit. J. exper. Biol. **5**, 102—111.

Greenstein, J. P., 1954: Biochemistry of cancer. 2nd edition. New York.

Günther, E., 1957: Der Einfluß chemischer und physikalischer Faktoren auf die protoplasmatischen Eigenschaften isolierter *Elodea*-Blätter unter besonderer Berücksichtigung der Zusammenhänge zwischen Wachstumsphase und Plasmaviskosität. Protoplasma **48**, 1—52.

Haddow, A., 1935: Influence of certain polycyclic hydrocarbons on the growth of the Jensen rat sarcoma. Nature **136**, 868—869.

— 1951: Advances in the study of chemical carcinogenesis. Proc. roy. Soc. Med., Lond. **44**, 263—266.

Haeckel, E., 1857: Über die Gewebe des Flußkrebses. Arch. Anat. usw. 1857, 469—568.

Harding, C. V., 1949: Colloidal properties of nucleus. I. Effect of temperature on nuclear viscosity in the starfish egg. Proc. Soc. exper. Biol. a. Med. (Am.) **70**, 705—708.

HARDING, C. V., The effect of ultraviolet light on starfish egg protoplasm. Pubblic. Staz. Zool. Napoli 27, 318—330.
— and L. J. THOMAS, 1950: Ultraviolet light induced delay in cleavage of centrifuged *Arbacia* eggs. J. cellul. a. comp. Physiol. (Am.) 35, 403—411.
HARDING, D., 1951: Initiation of cell division in the *Arbacia* egg by injury substances. Physiol. Zool. 24, 54—69.
HARDY, W. B., 1895: On some histological and physiological features of the postoesophageal nerve cord of the crustacea. Phil. Trans. Roy. Soc. B 185, 83—117.
HARRIS, J. E., 1935: Studies on living protoplasm. I. Streaming movements in the protoplasm of the egg of *Sabellaria alveolata* (L.). J. exper. Biol. 12, 65—79.
— 1939: Studies on living protoplasm. II. The physical structure of the nucleus of the echinoderm oocyte. J. exper. Biol. 16, 258—277.
HARVEY, E. B., 1956: The American *Arbacia* and other sea urchins. Princeton.
HARVEY, E. N., 1932 a: Physical and chemical constants of the egg of the sea urchin, *Arbacia punctulata*. Biol. Bull. (Am.) 62, 141—154.
— 1932 b: The microscope-centrifuge and some of its applications. J. Franklin Inst. 214, 1—23.
— and A. L. LOOMIS, 1930: A microscope-centrifuge. Science 72, 42—44.
— and D. A. MARSLAND, 1932: The tension at the surface of *Amoeba dubia* with direct observations on the movement of cytoplasmic particles at high centrifugal speeds. J. cellul. a. comp. Physiol. (Am.) 2, 75—97.
HEILBRONN A., 1914: Zustand des Plasmas und Reizbarkeit. Ein Beitrag zur Physiologie der lebenden Substanz. Jb. wiss. Bot. 54, 357—390.
— 1922: Eine neue Methode zur Bestimmung der Viskosität lebender Protoplasten. Jb. wiss. Bot. 61, 284—338.
HEILBRUNN, L. V., 1915: Studies in artificial parthenogenesis. II. Physical changes in the egg of *Arbacia*. Biol. Bull. (Am.) 29, 149—203.
— 1917: An experimental study of cell division. Anat. Rec. (Am.) 11, 487—489.
— 1920 a: An experimental study of cell-division. I. The physical conditions which determine the appearance of the spindle in sea-urchin eggs. J. exper. Zool. 30, 211—237.
— 1920 b: The physical effect of anesthetics upon living protoplasm. Biol. Bull. (Am.) 39, 307—315.
— 1921: Protoplasmic viscosity changes during mitosis. J. exper. Zool. 34, 417—447.
— 1923: The colloid chemistry of protoplasm. I. General considerations. II. The electrical charges of protoplasm. Amer. J. Physiol. 64, 481—498.
— 1924 a: The colloid chemistry of protoplasm. III. The viscosity of protoplasm at various temperatures. Amer. J. Physiol. 68, 645—648.
— 1924 b: The colloid chemistry of protoplasm. IV. The heat coagulation of protoplasm. Amer. J. Physiol. 69, 190—199.
— 1925 a: The electrical charges of living cells. Science 61, 236—237.
— 1925 b: The action of ether on protoplasm. Biol. Bull. (Am.) 49, 461—476.
— 1926 a: The centrifuge method of determining protoplasmic viscosity. J. exper. Zool. 43, 313—320.
— 1926 b: The absolute viscosity of protoplasm. J. exper. Zool. 44, 255—278.
— 1927: The colloid chemistry of protoplasm. V. A preliminary study of the surface precipitation reaction of living cells. Arch. exper. Zellforsch. 4, 246—263.
— 1928: The Colloid Chemistry of Protoplasm. Berlin.
— 1929 a: The absolute viscosity of *Amoeba* protoplasm. Protoplasma 8, 65—69.
— 1929 b: Protoplasmic viscosity of *Amoeba* at different temperatures. Protoplasma 8, 58—64.
— 1934: The effect of anesthetics on the surface precipitation reaction. Biol. Bull. (Am.) 66, 264—275.
— 1937: An Outline of General Physiology. 1st edition, Philadelphia.
— 1940: The action of calcium on muscle protoplasm. Physiol. Zool. 13, 88—94.
— 1943: An Outline of General Physiology. 2nd edition. Philadelphia.
— 1950: Viscosity measurements. Chap. 4 in Uber's Biophysical Research Methods. New York.
— 1951: The colloid chemistry of narcosis. In: Mécanisme de la Narcose. Colloq. internat. centre nat. recherche sci. (Paris) No. 26, 163—177.
— 1952: An Outline of General Physiology. 3rd edition. Philadelphia.
— 1956 a: The Dynamics of Living Protoplasm. New York.
— 1956 b: The aging of *Chaetopterus* eggs. Protoplasma 46, 317—321.
— 1956 c: Cellular physiology and aging. Fed. Proc. 15, 948—953.

Heilbrunn, L. V., A. B. Chaet, A. Dunn, and W. L. Wilson, 1954: Antimitotic substances from ovaries. Biol Bull. (Am.) **106**, 158—168.
— and K. Daugherty, 1931: The action of the chlorides of sodium, potassium, calcium, and magnesium on the protoplasm of *Amoeba dubia*. Physiol. Zool. **4**, 635—651.
— — 1933: The action of ultraviolet rays on protoplasm. Protoplasma **18**, 596—619.
— — 1934: A further study of the action of potassium on *Amoeba* protoplasm. J. cellul. a. comp. Physiol. (Am.) **5**, 207—218.
— — 1939: The electric charge of protoplasmic colloids. Physiol. Zool. **12**, 1—12.
— D. L. Harris, P. G. LeFevre, W. L. Wilson, and A. A. Woodward, 1946: Heat death, heat injury, and toxic factor. Physiol. Zool. **19**, 404—429.
— and D. Mazia, 1936: The action of radiations on living protoplasm. In Duggar's Biological Effects of Radiation, New York 1936, vol. 1, pp. 625—676.
— and F. J. Wiercinski, 1947: The action of various cations on muscle protoplasm. J. cellul. a. comp. Physiol. (Am.) **29**, 15—32.
— and K. M. Wilbur, 1937: Stimulation and nuclear breakdown in the *Nereis* egg. Biol. Bull. (Am.) **73**, 557—564.
— and W. L. Wilson, 1948: Protoplasmic viscosity changes during mitosis in the egg of *Chaetopterus*. Biol. Bull. (Am.) **95**, 57—68.
— — 1949: The effect of heparin on cell division. Proc. Soc. exper. Biol. a. Med. (Am.) **70**, 179—182.
— — 1950 a: Effect of bacterial polysaccharide on cell division. Science **112**, 56—57.
— — 1950 b: The prevention of cell division by anticlotting agents. Protoplasma **39**, 389—399.
— — 1955 a: The initiation of cell division in the *Chaetopterus* egg. Protoplasma **44**, 377—384.
— — 1955 b: Changes in the protoplasm during maturation. Biol. Bull. (Am.) **109**, 271—275.
— — 1956: Antimitotic substances from the ovaries of vertebrates. Biol. Bull. (Am.) **110**, 153—156.
— — 1957: A rational approach to the problem of cancer chemotherapy. Biol. Bull. In press.
— — and D. Harding, 1951: The action of tissue extracts on cell division. J. nat. Cancer Inst. **11**, 1287—1298.
— — T. R. Tosteson, E. Davidson, and R. J. Rutman, 1957: The antimitotic and carcinostatic action of ovarian extracts. Biol. Bull. (Am.) **113**, 129—134.
— and R. A. Young, 1930: The action of ultra-violet rays on *Arbacia* egg protoplasm. Physiol. Zool. **3**, 330—341.
Hellerman, L., 1937: Reversible inactivations of certain hydrolytic enzymes. Physiol. Rev. (Am.) **17**, 454—484.
Helmholtz, Arminius (Hermann), 1842: De fabrica systematis nervosi evertebratorum. Diss. Berlin.
Hermann, S., 1932: Die Beeinflußbarkeit der Zustandsform des Calciums im Organismus durch Adrenalin. Arch. exper. Path. (D.) **167**, 82—84.
Herrick, F. H., 1895: Movements of the nucleolus through the action of gravity. Anat. Anz. **10**, 337—340.
Hertwig, O., 1909: Allgemeine Biologie. 3rd ed. Jena.
Hiramoto, Y., 1956: Physical state of muscle protoplasm. Annot. Zool. Japon. **29**, 63—68.
Hofmeister, W., 1867: Die Lehre von der Pflanzenzelle. Leipzig.
Howard, E., 1932: The structure of protoplasm as indicated by a study of the apparent viscosity of sea-urchin eggs at various shearing forces. J. cellul. a. comp. Physiol (Am.) **1**, 355—369.
Hunter, F. E., and O. H. Lowry, 1956: The effects of drugs on enzyme systems. Pharmacol. Rev. **8**, 89—135.
Jacobs, M. H., 1920: The production of intracellular acidity by neutral and alkaline solutions containing carbon dioxide. Amer. J. Physiol. **53**, 457—463.
— 1922 a: The effects of carbon dioxide on the consistency of protoplasm. Biol. Bull. (Am.) **42**, 14—30.
— 1922 b: The influence of ammonium salts on cell reaction. J. gen. Physiol (Am.) **5**, 181—188.
Josing, E., 1901: Der Einfluß der Außenbedingungen auf die Abhängigkeit der Protoplasmaströmung vom Licht. Jb. wiss. Bot. **36**, 197—228.

JUNGERS, V., 1934: Die Verlagerungsfähigkeit des Zellinhaltes der Zwiebelschuppen von *Allium cepa* durch Zentrifugieren. Protoplasma 21, 351—361.

KAMIYA, N., 1940: The control of protoplasmic streaming. Science 92, 462—463.

— 1942: Physical aspects of protoplasmic streaming. In Seifriz' A Symposium on the Structure of Protoplasm. Pp. 199—244. Ames, Iowa.

KELLY, J. W., 1950: The localization of a metachromatic substance in the *Chaetopterus* egg. Protoplasma 39, 386—388.

— 1951: Effects of x-ray and trypsin on the metachromasy of heparin-basic protein combinations. Biol. Bull. (Am.) 101, 223.

— 1954: Metachromasy in the eggs of fifteen lower animals. Protoplasma 43, 329—346.

KESSLER, W., 1935: Über die inneren Ursachen der Kälteresistenz der Pflanzen. Planta 24, 312—352.

KIMBALL, R. F., 1955: The effects of radiation on protozoa and the eggs of invertebrates other than insects. In Hollaender's Radiation Biology, vol. 2, New York 1955, pp. 285—331.

KLEMM, P., 1895: Desorganisationserscheinungen der Zelle. Jb. wiss. Bot. 28, 627—700.

KRIEGER, K. R., 1880: Über das Centralnervensystem des Flußkrebses. Z. wiss. Zool. 33, 527—594.

KRISZAT, G., 1950: Die Wirkung von Adenosintriphosphat und Calcium auf Amöben (*Chaos chaos*). Arkiv för Zoologi, ser. 2, vol. 1, 477—485.

KÜHNE, W., 1863: Eine lebende Nematode in einer lebenden Muskelfaser beobachtet. Virchow's Arch. 26, 222—224.

— 1864: Untersuchungen über das Protoplasma und die Contractilität. Leipzig.

LADENBURG, R., 1907: Über den Einfluß von Wänden auf die Bewegung einer Kugel in einer reibenden Flüssigkeit. Ann. Physik (4) 23, 447—458.

LAWACZEK, H., 1928: Über das Verhalten des Kalziums unter Adrenalin. Dtsch. Arch. klin. Med. 160, 309—322.

LEVINE, R., and M. S. GOLDSTEIN, 1955: On the mechanism of action of insulin. In: Recent Progress in Hormone Research, edited by G. Pincus, pp. 343—380. New York.

LIBBRECHT, W., 1920: L'adrénaline et ses rapports avec les ions K et Ca. Arch. internat. Phys. 15, 352—360.

LILLIE, F. R., 1908: Polarity and bilaterality of the annelid egg. Experiments with centrifugal force. Biol. Bull. (Am.) 16, 54—79.

LOEB, J., 1901: Experiments on artificial parthenogenesis in annelids (*Chaetopterus*) and the nature of the process of fertilization. Amer. J. Physiol. 4, 423—439.

— 1913: Artificial Parthenogenesis and Fertilization. Chicago.

LOICQ, R., 1937: Recherches sur les effets de la colchicine sur la coagulation du sang. Arch. internat. Méd. expér. 12, 371—396.

LOPRIORE, G., 1897: Azione dei raggi X sul protoplasma della cellula vegetale vivente. Nuova Rassegna.

LORAND, L., 1954: Interaction of thrombin and fibrinogen. Physiol. Rev. (Am.) 34, 742—752.

MACDOUGALD, T. J., H. W. BEAMS, and R. L. KING, 1937: Growth of ultracentrifuged cells in tissue culture. Proc. Soc. exper. Biol. a. Med. (Am.) 37, 234—235.

MAST, S. O., 1926: Structure, movement, locomotion, and stimulation in *Amoeba*. J. Morph. (Am.) 41, 347—425.

— and W. L. DOYLE, 1935: Structure, origin and function of cytoplasmic constituents in *Amoeba proteus* with special reference to mitochondria and Golgi substance. II. Origin and function based on experimental evidence; effect of centrifuging on *Amoeba proteus*. Arch. Protistenk. 86, 278—306.

MAXIMOW, N. A., and L. V. MOZHAEVA, 1944: Age variations in the colloid-chemical properties of protoplasm of vegetable cells. II. Variations in permeability and viscosity in the leaf cells of broad-beans and oats. C. r. (Doklady) Acad. Sci. URSS 42, 277—280.

MAZIA, D., 1954 a: SH and growth. In: Glutathione—A Symposium, edited by S. COLOWICK and others, New York, pp. 209—228.

— 1954 b: The particulate organization of the chromosome. Proc. nat. Acad. Sci. 40, 521—527.

Mazia, D., and J. M. Clark, 1936: Free calcium in the action of stimulating agents on *Elodea* cells. Biol. Bull. (Am.) **71**, 306—323.

McNamara, B., S. Krop, and E. A. McKay, 1948: The effect of calcium on the cardiovascular stimulation produced by acetylcholine. J. Pharmacol. (Am.) **92**, 153—161.

Meyer, A., 1921: Analyse der Zelle der Pflanzen und Tiere. Zweiter Teil. Jena.

Miehe, H., 1901: Über die Wanderungen des pflanzlichen Zellkernes. Flora **88**, 105—142.

Moder, A., 1932: Beiträge zur protoplasmatischen Anatomie des *Helodea*-Blattes. Protoplasma **16**, 1—55.

Moser, F., 1939: Studies on a cortical layer response to stimulating agents in the *Arbacia* egg. I. Response to stimulation. J. exper. Zool. **80**, 423—445.

Most, S., 1951: Effect of Shear's polysaccharide on plasma clotting. Nature **168**, 342—343.

Murphy, Q., 1940: The effect of temperature on the viscosity of *Amoeba dubia* protoplasm. J. cellul. a. comp. Physiol. (Am.) **16**, 401—404.

Northen, H. T., 1936: Is protoplasm elastic? Bot. Gaz. **98**, 421—424.

— 1938 a: Studies on protoplasmic structure in *Spirogyra*. I. Elasticity. Protoplasma **31**, 1—8.

— 1938 b: Protoplasmic structure in *Syirogyra*. III. Effects of anesthetics on protoplasmic elasticity. Bot. Gaz. **100**, 238—244.

— 1939: Studies on the protoplasmic nature of stimulation and anesthesia. Cytologia **10**, 105—112.

-- 1940: Studies on the protoplasmic nature of stimulation and anaesthesia. II. Plant Physiol. **15**, 645—660.

— 1942: Relationship of dissociation of cellular proteins by auxins to growth. Bot. Gaz. **103**, 668—683.

— 1950: Alterations in the structural viscosity of protoplasm by colchicine and their relationship to C-mitosis and C-tumor formation. Amer. J. Bot. **37**, 705—711.

— and R. MacVicar, 1939: Studies of protoplasmic structure in *Spirogyra*. VI. Effects of sound and electricity on elasticity. Cytologia **10**, 18—22.

— and R. T. Northen, 1938: Studies of protoplasmic structure in *Spirogyra*. II. Alterations of protoplasmic elasticity. Protoplasma **31**, 9—19.

Otis, A. B., 1942: Effects of certain drugs and ions on the oyster heart. Physiol. Zool. **15**, 418—435.

Overbeek, J. T. G., 1952: Rheology of lyophobic systems. In Kruyt's Colloid Science, vol. 1, pp. 342—368. New York.

— and H. G. Bungenberg de Jong, 1949: Sols of macromolecular colloids with electrolytic nature. In Kruyt's Colloid Science, vol. 2, pp. 184—231. New York.

Pantin, C. F. A., 1924: Temperature and the viscosity of protoplasm. J. Marine Biol. Ass. **13**, 331—339.

Pekarek, J., 1930 a: Absolute Viskositätsmessung mit Hilfe der Brownschen Molekularbewegung. I. Prinzip der Methode, Voraussetzungen, Fehlerquellen der Messungen. Protoplasma **10**, 510—532.

— 1930 b: Absolute Viskositätsmessung mit Hilfe der Brownschen Molekularbewegung. Viskositätsbestimmung des Zellsaftes der Epidermiszellen von *Allium cepa* und des Amoeben-Protoplasmas. Protoplasma **11**, 19—48.

— 1931: Absolute Viskositätsmessungen mit Hilfe der Brownschen Molekularbewegung. III. a) Viskositätsmessungen an destilliertem Wasser. b) Viskositätsmessungen an Glyzerin-Wasser-Gemischen. c) Viskositätsmessungen des Zellsaftes der Protonemazellen von *Leptobryum piriforme*. Protoplasma **13**, 637—665.

— 1932: Absolute Viskositätsmessungen mit Hilfe der Brownschen Molekularbewegung. IV. Plasmaviskositätsmessungen an Rhizoiden von *Chara fragilis* Desv. Protoplasma **17**, 1—24.

— 1933: Absolute Viskositätsmessungen mit Hilfe der Brownschen Molekularbewegung. VII. Der Einfluß des Lichtes auf die Zellsaftviskosität. Protoplasma **20**, 359—375.

Pfeiffer, H., 1939: Vervollkommnung der rheologischen Versuchseinrichtung für Protoplasmatropfen. Protoplasma **33**, 311—313.

— 1940: Rheologische, polarisationsoptische und beugungspolarisatorische Untersuchungen an Protoplasmatropfen und einigen Modellsubstanzen. Protoplasma **34**, 347—352.

Punt, A., 1942: L'influence de la température et des électrolytes sur la contracture acétylcholinique. Arch. néerl. de physiol. 26, 212—221.

Purr, A., 1935: The activation phenomena of papain and cathepsin. Biochem. J. (Brit.) 29, 13—20.

Rapkine, L., 1931: Sur les processus chimiques au cours de la division cellulaire. Ann. physiol. physicochim. biol. 7, 382—418.

Rayleigh, Lord, 1893: On the flow of viscous liquids, especially in two dimensions. Phil. Mag. (5) 36, 354—372.

Rieser, P., 1949 a: The protoplasmic viscosity of muscle. Protoplasma 39, 95—98.
— 1949 b: The protoplasmic viscosity of muscle and nerve. Biol. Bull. (Am.) 97, 245—246.
— 1955 a: Effect of roentgen irradiation and other types of injurious agents on capillaries. Proc. Soc. exper. Biol. a. Med. (Am.) 89, 39—41.
— 1955 b: The effect of roentgen rays on the colloidal properties of the starfish egg. Biol. Bull. (Am.) 109, 108—112.
— and A. M. Kaye, 1953: Release of an anticoagulant from irradiated Spisula eggs. Biol. Bull. (Am.) 105, 380.

Robbie, W. A., 1949: Respiration of the tissues of some invertebrates and its inhibition by cyanide. J. gen. Physiol. (Am.) 32, 655—670.

Rochlin-Gleichgewicht, E. J., 1930: Effect of radon on chlorophyll containing cells. Vestnik Roentgen. i Radio 8, 387—407.

Röder, H. L., 1939: Rheology of suspensions: a study of dilatancy and thixotropy. Thesis, Utrecht. Published in Amsterdam.

Ruge, U., 1940: Kritische zell- und entwicklungsphysiologische Untersuchungen an den Blattzähnen von Helodea densa. Flora N. F. 34, 311—376.

Runnström, J., 1928 a: Die Veränderungen der Plasmakolloide bei der Entwicklungserregung des Seeigeleies. Protoplasma 4, 388—514.
— 1928 b: Über die Veränderung der Plasmakolloide bei der Entwicklungserregung des Seeigeleies. Protoplasma 5, 201—310.
— and G. Kriszat, 1950: On the effect of adenosine triphosphoric acid and of Ca on the cytoplasm of the egg of the sea urchin, Psammechinus miliaris. Exper. Cell Res. 1, 284—303.

Ruppert, W., 1924: Empfindlichkeitsänderungen des Ascaris-Eies auf verschiedenen Stadien der Entwicklung gegenüber der Einwirkung ultravioletter Strahlen. Z. wiss. Zool. 123, 103—155.

Sato, H., M. Belkin, and E. Essner, 1956: Effect of Nitromin on mitosis and cytoplasmic volume in the cells of two mouse ascites tumors. J. nat. Cancer Inst. 17, 421—433.

Schiller, L., 1932: Fallversuche mit Kugeln und Scheiben. In: Handbuch der Experimentalphysik, edited by W. Wien and F. Harms. Leipzig, vol. 4, part 2, pp. 337—387.

Schleip, W., 1923: Die Wirkung des ultravioletten Lichtes auf die morphologischen Bestandteile des Ascaris-Eies. Arch. Zellforsch. 17, 288—367.

Schmidt, H., 1939: Plasmazustand und Wasserhaushalt bei Lamium maculatum. Protoplasma 33, 25—43.
— K. Diwald und O. Stocker, 1940: Plasmatische Untersuchungen an dürreempfindlichen und dürreresistenten Sorten landwirtschaftlicher Kulturpflanzen. Planta 31, 559—596.

Scott Blair, G. W., 1938: An introduction to industrial rheology. Philadelphia.
— 1949: A survey of general and applied rheology. London.

Shear, M. J., and A. Perrault, 1944: Chemical treatment of tumors. IX. Reactions of mice with primary subcutaneous tumors to injection of a hemorrhage-producing bacterial polysaccharide. J. nat. Cancer Inst. 4, 461—476.

Shirley, E. S., and H. E. Finley, 1949: The effects of ultra-violet radiations on Spirostomum. ambiguum. Trans. amer. Microsc. Soc. 68, 136—153.

Simha, R., 1949: Effect of concentration on the viscosity of dilute solutions. J. Res. Nat. Bur. Standards 42, 409—418.
— 1950: The concentration dependence of viscosities in dilute solutions. J. Colloid Sci. 5, 386—392.
— 1952: A treatment of the viscosity of concentrated suspensions. J. Appl. Physics 23, 1020—1024.

Smitten, N. A., 1946: Vital studies of the neuroplasm. Amer. Rev. Soviet Med. 3, 414—425.

Smoluchowski, M. von, 1906: Zur kinetischen Theorie der Brownschen Molekular-bewegung und der Suspensionen. Ann. Physik (4) 21, 756—780.
— 1916: Theoretische Bemerkungen über die Viskosität der Kolloide. Kolloid-Z. 18, 190—195.
Stålfelt, M. G., 1946: The influence of light upon the viscosity of protoplasm. Arkiv för Bot. 33 A, No. 4, 1—17.
— 1947: Effect of heteroauxin and colchicine on protoplasmic viscosity. Proc. 6th Internat. Congress Exp. Cytology 63—78. Also Exper. Cell. Res. 1 Supp. 63—78, 1949.
— 1954: The relation between the fluidity of the protoplasm and the insertion and function of the leaves. Physiol. Plantarum 7, 354—374.
— 1956: Viscosität. In Ruhland's Handbuch der Pflanzenphysiologie. Vol. 2, pp. 591—606. Berlin.
— 1957: The influence of distilled water on the fluidity of protoplasm. Protoplasma 48, 134—142.
Strugger, S., 1934: Beiträge zur Physiologie des Wachstums. I. Zur protoplasma-physiologischen Kausalanalyse des Streckungswachstums. Jb. wiss. Bot. 79, 406—471.
Sugai, K., 1944: Studies on protease. VI. The effect of the addition of various salts on the tryptic activity. J. Biochem. (Japan) 36, 91—100.
Swann, M. 1956: Review of the Dynamics of Living Protoplasm by L. V. Heilbrunn, Quart. J. exper. Physiol. 41, 480—481.
Szent-Györgyi, A., 1949: Free-energy relations and contraction of actomyosin. Biol. Bull. (Am.) 96, 140—161.
Thomas, L. J., 1951: A blood anticoagulant from surf clams. Biol. Bull. (Am.) 101, 230—231.
— 1954: The localization of heparin-like blood anticoagulant substances in the tissues of Spisula solidissima. Biol. Bull. (Am.) 106, 129—138.
Thornton, F. E., 1935: The action of sodium, potassium, calcium and magnesium ions on the plasmagel of Amoeba proteus at different temperatures. Physiol. Zool. 8, 246—254.
Umrath, K., 1942: Ausbreitung der durch Verwundung bedingten Viskositäts-verminderung bei Spirogyra. Protoplasma 36, 410—413.
Ungar, G., and E. Damgaard, 1954: Protein breakdown in thermal injury. Proc. Soc. exper. Biol. a. Med. (Am.) 87, 378—383.
Uretz, R. B., and R. E. Zirkle, 1955: Disappearance of spindles in sand-dollar blastomeres after ultraviolet irradiation of cytoplasm. Biol. Bull. (Am.) 109, 370.
Van Herwerden, M. A., 1925: Reversible Gelbildung in Epithelzellen der Frosch-larve und ihre Anwendung zur Prüfung auf Permeabilitätsunterschiede in der lebenden Zelle. Arch. exper. Zellforsch. 1, 145—159.
Virgin, H. I., 1951: The effect of light on the protoplasmic viscosity. Physiol. Plantarum 4, 255—357.
— 1953: Physical properties of protoplasm. Ann. Rev. Plant Physiol. 4, 363—382.
— and L. Ehrenberg, 1953: Effects of β- and γ-rays on the protoplasmic viscosity of Helodea cells. Physiol. Plantarum 6, 159—165.
Voerkel, S. H., 1933: Untersuchungen über die Phototaxis der Chloroplasten. Planta 21, 156—205.
Weber, F., 1917: Die Plasmaviskosität pflanzlicher Zellen. Z. allg. Physiol. 18, 1—20.
— 1921: Zentrifugierungsversuche mit ätherisierten Spirogyren. Biochem. Z. 126, 21—32.
— 1922: Reversible Viskositätserhöhung bei Narkose. Ber. dtsch. bot. Ges. 40, 212—216.
— 1924 a: Methoden der Viskositätsbestimmung des lebenden Protoplasmas. In: Abderhalden's Handb. d. biol. Arbeitsmethoden, Abt. 11, Teil 2, 655—718.
— 1924 b: Krampf-Plasmolyse bei Spirogyra. Arch. ges. Physiol. 206, 629—634.
— 1924 c: Plasmolyseform und Protoplasmaviskosität. Öst. bot. Z. 73, 261—266.
— 1925 a: Plasmolyseform und Ätherwirkung. Arch. ges. Physiol. 208, 705—717.
— 1925 b: Schrauben-Plasmolyse bei Spirogyra. Ber. dtsch. bot. Ges. 43, 217—223.
— 1925 c: Plasmolyseform und Kernform funktionierender Schließzellen. Jb. wiss. Bot. 64, 687—701.
— 1925 d: Über die Beurteilung der Plasmaviskosität nach der Plasmolyseform (Untersuchungen an Spirogyra). Z. wiss. Mikrosk. 42, 146—156.

WEBER, F., 1925 e: Physiologische Ungleichheit bei morphologischer Gleichheit. Öst. bot. Z. **74**, 256—261.
— 1927: Cytoplasma- und Kern-Zustandsänderungen bei Schließzellen. Protoplasma **2**, 305—311.
— 1929 a: Plasmolyse-Zeit-Methode. Protoplasma **5**, 622—624.
— 1929 b: Plasmolysezeit und Lichtwirkung. Protoplasma **7**, 256—258.
— and G. WEBER, 1916 (1917): Die Temperaturabhängigkeit der Plasmaviskosität. Ber. dtsch. bot. Ges. **34**, 836—846.
WEIMAR, V., 1953: Calcium binding in frog muscle brei. Physiol. Zool. **26**, 231—242.
— 1955 a: Trypsin shock in frogs. Amer. J. Physiol. **182**, 351—358.
— 1955 b: Activation of proteolytic enzymes in peptone shock in frogs. Arch. internat. Pharmacodynam. **103**, 419—434.
WENSE, T., 1934: Untersuchungen über die Einwirkung von Adrenalin auf Paramecien. Ein Beitrag zur Frage der kolloid-chemischen Wirkung des Adrenalins auf das Protoplasma. Arch. exper. Path. (D.) **176**, 49—58.
WHITING, A. R., 1950: Absence of mutagenic action of X-rayed cytoplasm in *Habrobracon*. Proc. nat. Acad. Sci. **36**, 368—372.
WIERCINSKI, F. J., 1955: The pH of animal cells. Protoplasmatologia B 2 C.
— and B. A. COOKSON, 1949: Action of dilute trypsin on muscle protoplasm. Fed. Proc. **8**, 165.
WILBUR, K. M., 1940: Effects of colchicine upon viscosity of the *Arbacia* egg. Proc. Soc. exper. Biol. a. Med. (Am.) **45**, 696—700.
WILLIAMS, M., 1923: Observations on the action of X-rays on plant cells. Ann. Bot. **37**, 217—223.
— 1925: Some observations on the action of radium on certain plant cells. Ann. Bot. **39**, 547—562.
WILLSTÄTTER, R., W. GRASSMANN, und O. AMBROS, 1926: Blausäure-Aktivierung und -Hemmung pflanzlicher Proteasen. Zweite Abhandlung über pflanzliche Proteasen. Z. physiol. Chem. **151**, 286—306.
WILSON, W. L., 1950: The effect of Roentgen rays on protoplasmic viscosity changes during mitosis. Protoplasma **39**, 305—317.
— and L. V. HEILBRUNN, 1952: The protoplasmic cortex in relation to stimulation. Biol. Bull. (Am.) **103**, 139—144.
— — 1957: The relation of protoplasmic gelation to oxidative processes. Exper. Cell Research. **13**, 234—243.
— and Y. SCHUB, 1953: The effect of strong centrifugal force on the development of *Chaetopterus* eggs. Biol. Bull. (Am.) **105**, 389.
WOODWARD, A. A., 1948: Protoplasmic clotting in isolated muscle fibers. J. cellul. a. comp. Physiol. (Am.) **31**, 359—394.
YOUNG, J. Z., 1936: The structure of nerve fibers in cephalopods and crustacea. Proc. roy. Soc., Lond. **B 121**, 319—337.
YUNG, E., 1878: Recherches sur la structure intime et les fonctions du système nerveux central chez les crustacés décapodes. Arch. Zool. exptl. et gén. **7**, 401—534.
ZAKRZEWSKI, Z., 1932: Untersuchungen über den Einfluß von Heparin auf das Wachstum von transplantablen Sarkomen. Bull. internat. acad. polon. sci. Classe méd. **1932**, 239—259.

SPRINGER-VERLAG IN WIEN

Protoplasmatologia. Handbuch der Protoplasmaforschung.

Herausgegeben von **L. V. Heilbrunn**, Philadelphia, und **F. Weber**, Graz.

Das Handbuch erscheint in selbständigen Einzelveröffentlichungen, die zu Bänden vereinigt werden. Jeder selbständig erscheinende Handbuchteil ist einzeln käuflich. Bei Verpflichtung zur Abnahme des gesamten Handbuches, bei Vorbestellung der einzelnen Teile sowie für Abonnenten der Zeitschrift „Protoplasma" ermäßigt sich der Preis um 20%. Über die Disposition des Gesamtwerkes und die nächsten Veröffentlichungen gibt der Verlag bereitwilligst Auskunft.

Zuletzt erschienen:

Biocolloids and Their Interactions with Special Reference to Coacervates and Related Systems. By H. L. Booij and H. G. Bungenberg de Jong, Department of Medical Chemistry, University of Leiden. With 159 figures. IV, 162 pages. Gr.-8°. 1956. **Band I. Grundlagen. 2.**
S 312.—, DM 52.—, sfr. 53.30, $ 12.40

Cytoplasmastruktur in Pflanzenzellen. Von Lotte Reuter, Pflanzenphysiologisches Institut der Universität Wien. Mit 27 Textabbildungen. IV, 44 Seiten. — **Intravakuoläres Protoplasma.** Von Ernst Küster †, Gießen (Lahn). Mit 7 Textabbildungen. 12 Seiten. — **Plasmodesmata (Vegetable Kingdom).** By A. D. J. Meeuse, Pretoria. With 14 figures. 43 pages. Gr.-8°. 1957. Band II. Cytoplasma. A. Morphologie. 1. Mikroskopische Morphologie. a, b, c.
S 186.—, DM 31.—, sfr. 31.70, $ 7.40

Die Ascorbinsäure in der Pflanzenzelle. Von Helmut Metzner, Göttingen. Mit 26 Textabbildungen. IV, 68 Seiten. — **Vitamin C in the Animal Cell.** By G. H. Bourne, Department of Histology, London Hospital Medical College, London E. 1. With 49 figures. 91 pages. Gr.-8°. 1957. Band II. Cytoplasma. B. Chemie. 2. Spezielle Cytochemie und Histochemie. b. Organische Verbindungen. α.
S 300.—, DM 50.—, sfr. 51.20, $ 11.90

The Metachromatic Reaction. By John W. Kelly, Department of Anatomy, Medical College of Virginia, Richmond, Virginia. With 5 figures. IV, 98 pages. Gr.-8°. 1956. Band II. Cytoplasma. D. Vitalfärbung. Vitalfluorochromierung. 2.
S 210.—, DM 35.—, sfr. 35.90, $ 8.35

Osmotischer Wert, Saugkraft, Turgor. Von G. Blum, Botanisches Institut der Universität Freiburg/Schweiz. Mit 12 Textabbildungen. IV, 98 Seiten. — **Plasmoptyse.** Von Ernst Küster †, Botanisches Institut der Universität Gießen. Mit 10 Textabbildungen. 39 Seiten. — **Plasmorrhyse.** Von Hans H. Pfeiffer, Laboratorium für Polarisations-Mikroskopie, Bremen. Mit 8 Textabbildungen. 16 Seiten. — **Plasmoschise.** Von Hans H. Pfeiffer, Laboratorium für Polarisations-Mikroskopie, Bremen. Mit 1 Textabbildung. 7 Seiten. Gr.-8°. 1958. Band II. Cytoplasma. C. Physik, Physikalische Chemie, Kolloidchemie. 7. Osmotische Zustandsgrößen. a. b. c. d.
S 300.—, DM 50.—, sfr. 51.20, $ 11.90

Le vacuome de la cellule végétale. Morphologie. Par Pierre Dangeard, Laboratoire de Botanique, Faculté des Sciences, Université de Bordeaux. Avec 26 figures. IV, 41 pages. — **Le vacuome animal.** Par Raymond Hovasse, Faculté des Sciences de Clermont-Ferrand, France. Avec 16 figures. 37 pages. — **Contractile Vacuoles of Protozoa.** By J. A. Kitching, Department of Zoology, University of Bristol. With 20 figures. 45 pages. — **Food Vacuoles.** By J. A. Kitching, Department of Zoology, University of Bristol. With 24 figures. 54 pages. Gr.-8°. 1956. Band III. Cytoplasma-Organellen. D. Vacuome. 1. 2. 3 a. 3 b.
S 402.—, DM 67.—, sfr. 68.80, $ 16.—

Effets biologiques des Radiations. Aspects biochimiques. Par Maurice Errera, Laboratoire de Morphologie animale, Laboratoire de recherches pour la protection des Populations civiles, Faculté des Sciences, Université Libre de Bruxelles. Avec 27 figures. IV, 241 pages. Gr.-8°. 1957. Band X. Pathologie des Protoplasmas. 3.
S 426.—, DM 71.—, sfr. 72.70, $ 16.90

Zu beziehen durch Ihre Buchhandlung